国外电子信息精品著作（影印版）

Digital System Design with SystemVerilog

SystemVerilog 数字系统设计

Mark Zwolinski

科学出版社

北京

图字:01-2012-2732

内 容 简 介

SystemVerilog是21世纪电子设计师必须掌握的最重要的语言之一，因为它是设计和验证复杂电子系统核心芯片的重要手段。本书是用SystemVerilog语言设计并验证数字系统的基本概念和具体方法。在介绍基本语法的基础上，阐述了如何用SystemVerilog构成数字电路、组件和系统，以及应该如何使用SystemVerilog搭建测试平台，并对设计进行验证。

本书既适合作电子、自动化和计算机专业本科生和研究生的教科书，也适合已经掌握Verilog和VHDL硬件描述语言的工程师使用。

图书在版编目(CIP)数据

SystemVerilog数字系统设计=Digital System Design with SystemVerilog:英文/(美)马克(Mark Z.)编著. 一影印版. 一北京:科学出版社，2012

(国外电子信息精品著作)

ISBN:978-7-03-034380-2

Ⅰ.S… Ⅱ.马… Ⅲ.数字系统-系统设计-英文 Ⅳ.TP271

中国版本图书馆CIP数据核字(2012)第103154号

责任编辑:张宇 佘丁 / 责任印制:张倩 / 封面设计:陈敬

科学出版社 出版

北京东黄城根北街16号

邮政编码: 100717

http://www.sciencep.com

源海印刷有限责任公司 印刷

科学出版社发行 各地新华书店经销

*

2012年6月第 一 版 开本: 787×1092 1/16

2012年6月第一次印刷 印张: 25

字数: 551 000

定价: 80.00 元

(如有印装质量问题，我社负责调换)

《国外电子信息精品著作》序

20 世纪 90 年代以来，信息科学技术成为世界经济的中坚力量。随着经济全球化的进一步发展，以微电子、计算机、通信和网络技术为代表的信息技术，成为人类社会进步过程中发展最快、渗透性最强、应用面最广的关键技术。信息技术的发展带动了微电子、计算机、通信、网络、超导等产业的发展，促进了生命科学、新材料、能源、航空航天等高新技术产业的成长。信息产业的发展水平不仅是社会物质生产、文化进步的基本要素和必备条件，也是衡量一个国家的综合国力、国际竞争力和发展水平的重要标志。在中国，信息产业在国民经济发展中占有举足轻重的地位，成为国民经济重要支柱产业。然而，中国的信息科学支持技术发展的力度不够，信息技术还处于比较落后的水平，因此，快速发展信息科学技术成为我国迫在眉睫的大事。

要使我国的信息技术更好地发展起来，需要科学工作者和工程技术人员付出艰辛的努力。此外，我们要从客观上为科学工作者和工程技术人员创造更有利于发展的环境，加强对信息技术的支持与投资力度，其中也包括与信息技术相关的图书出版工作。

从出版的角度考虑，除了较好较快地出版具有自主知识产权的成果外，引进国外的优秀出版物是大有裨益的。洋为中用，将国外的优秀著作引进到国内，促进最新的科技成就迅速转化为我们自己的智力成果，无疑是值得高度重视的。科学出版社引进一批国外知名出版社的优秀著作，使我国从事信息技术的广大科学工作者和工程技术人员能以较低的价格购买，对于推动我国信息技术领域的科研与教学是十分有益的事。

此次科学出版社在广泛征求专家意见的基础上，经过反复论证、仔细遴选，共引进了接近 30 本外版书，大体上可以分为两类，第一类是基础理论著作，第二类是工程应用方面的著作。所有的著作都涉及信息领域的最新成果，大多数是 2005 年后出版的，力求“层次高、内容新、参考性强”。在内容和形式上都体现了科学出版社一贯奉行的严谨作风。

当然，这批书只能涵盖信息科学技术的一部分，所以这项工作还应该继续下去。对于一些读者面较广、观点新颖、国内缺乏的好书还应该翻译成中文出版，这有利于知识更好更快地传播。同时，我也希望广大读者提出好的建议，以改进和完善丛书的出版工作。

总之，我对科学出版社引进外版书这一举措表示热烈的支持，并盼望这一工作取得更大的成绩。

王越

中国科学院院士

中国工程院院士

2006 年 12 月

Contents

List of Figures

List of Tables

Preface

About This Book

When *Digital System Design with VHDL* was published, the idea of combining a text on digital design with one on a hardware description language seemed novel. At about the same time, several other books with similar themes were published. *Digital System Design with VHDL* has now been adopted by several universities as a core text and has been translated into Polish, Chinese, Japanese, and Italian. I had thought about writing *Digital System Design with Verilog*, but I had (and still have) some doubts about using Verilog as a teaching language despite its widespread use. Soon after the second edition of *Digital System Design with VHDL* was published, a new hardware description language appeared—SystemVerilog. This new language removed many of my doubts about Verilog and even offered some noticeable advantages over VHDL. So the success of the first book and the appearance of the new language convinced me that the time had come for a new edition.

This book is intended as a student textbook for both undergraduate and postgraduate students. The majority of Verilog and SystemVerilog books are aimed at practicing engineers. Therefore, some features of SystemVerilog are not described at all in this book. Equally, aspects of digital design are covered that would not be included in a typical SystemVerilog book.

Syllabuses for electrical, electronic, and computer engineering degrees vary between countries and between universities or colleges. The material in this book has been developed over a number of years for second- and third-year undergraduates and for postgraduate students. It is assumed that students will be familiar with the principles of Boolean algebra and combinational logic design. At the University of Southampton, UK, the first-year undergraduate syllabus also includes introductions to synchronous sequential design and programmable logic. This book therefore builds upon these foundations. It has often been assumed that topics such as SystemVerilog are too specialized for second-year teaching and are best left to final year or postgraduate courses. There are several good reasons why SystemVerilog should be introduced earlier into the curriculum. With increasing integrated circuit complexity, there is a need for graduates with knowledge of SystemVerilog and the associated design tools. If left to the final year, there is little or no time for the student to apply such knowledge in project work. Second, conversations with colleagues from many countries suggest that today's students are opting for computer science or computer engineering courses in preference to electrical or electronic engineering. SystemVerilog offers a means to interest computing-oriented students in hardware design. Finally, simulation and synthesis tools and FPGA design kits are now mature and available relatively inexpensively to educational establishments on PC platforms.

Structure of This Book

Chapter 1 introduces the ideas behind this book, namely the use of electronic design automation tools and CMOS and programmable logic technology. We also consider some engineering problems, such as noise margins and fan-out. In Chapter 2, the principles of Boolean algebra and combinational logic design are reviewed. The important matter of timing and the associated problem of hazards are discussed. Some basic techniques for representing data are discussed.

SystemVerilog is introduced in Chapter 3 through basic logic gate models. The importance of documented code is emphasized. We show how to construct netlists of basic gates and how to model delays through gates. We also discuss parameterized models. The idea of using SystemVerilog to verify models with testbenches is introduced.

In Chapter 4, a variety of modeling techniques are described. Combinational building blocks, buffers, decoders, encoders, multiplexers, adders, and parity checkers are modeled using a range of concurrent and sequential SystemVerilog coding constructs. The SystemVerilog models of hardware introduced in this chapter and in Chapters 5, 6, and 7 are, in principle, synthesizable, although discussion of exactly what is supported is deferred until Chapter 10. Testbench design styles are again discussed in Chapter 4. In addition, the IEEE dependency notation is introduced.

Chapter 5 introduces various sequential building blocks: latches, flip-flops, registers, counters, memory, and a sequential multiplier. The same style as Chapter 4 is used, with IEEE dependency notation, testbench design, and the introduction of SystemVerilog coding constructs.

Chapter 6 is probably the most important chapter of the book and discusses what might be considered the cornerstone of digital design: the design of finite state machines. The ASM chart notation is used. The design process from ASM chart to

D flip-flops and next state and output logic is described. SystemVerilog models of state machines are introduced.

In Chapter 7, the concepts of the previous three chapters are combined. The ASM chart notation is extended to include coupled state machines and registered outputs, and hence to datapath-controller partitioning. From this, we explain the idea of instructions in hardware terms and go on to model a very basic microprocessor in SystemVerilog. This provides a vehicle to introduce interfaces and packages.

The design of testbenches is discussed in more detail in Chapter 8. After recapping the techniques given in earlier chapters, we go on to discuss testbench architecture, constrained random test generation, and assertion-based verification.

SystemVerilog remains primarily a modeling language. Chapter 9 describes the operation of a SystemVerilog simulator. The idea of event-driven simulation is first explained, and the specific features of a SystemVerilog simulator are then discussed.

The other, increasingly important, role of SystemVerilog is as a language for describing synthesis models, as discussed in Chapter 10. The dominant type of synthesis tool available today is for RTL synthesis. Such tools can infer the existence of flip-flops and latches from a SystemVerilog model. These constructs are described. Conversely, flip-flops can be created in error if the description is poorly written, and common pitfalls are described. The synthesis process can be controlled by constraints. Because these constraints are outside of the language, they are discussed in general terms. Suitable constructs for FPGA synthesis are discussed. Finally, behavioral synthesis, which promises to become an important design technology, is briefly examined.

Chapters 11 and 12 are devoted to the topics of testing and design for test. This area has often been neglected, but is now recognized as being an important part of the design process. In Chapter 11, the idea of fault modeling is introduced. This is followed by test generation methods. The efficacy of a test can be determined by fault simulation.

In Chapter 12, three important design-for-test principles are described: scan path, built-in self-test (BIST), and boundary scan. This has always been a very dry subject, but a SystemVerilog simulator can be used, for example, to show how a BIST structure can generate different signatures for fault-free and faulty circuits.

We use SystemVerilog as a tool for exploring anomalous behavior in asynchronous sequential circuits in Chapter 13. Although the predominant design style is currently synchronous, it is likely that digital systems will increasingly consist of synchronous circuits communicating asynchronously with each other. We introduce the concept of the fundamental mode and show how to analyze and design asynchronous circuits. We use SystemVerilog simulations to illustrate the problems of hazards, races, and setup and hold time violations. We also discuss the problem of metastability.

The final chapter introduces Verilog-AMS and mixed-signal modeling. Brief descriptions of digital-to-analog converters (DACs) and analog-to-digital converters (ADCs) are given. Verilog-AMS constructs to model such converters are given. We also introduce the idea of a phase-locked loop (PLL) here and give a simple mixed-signal model.

The Appendix briefly describes how SystemVerilog differs from earlier versions of Verilog.

At the end of each chapter a number of exercises have been included. These exercises are almost secondary to the implicit instruction in each chapter to simulate and, where appropriate, synthesize each SystemVerilog example. To perform these simulation and synthesis tasks, the reader may have to write his or her own testbenches and constraints files. The examples are available on the Web at zwolinski.org.

How to Use This Book

Obviously, this book can be used in a number of different ways, depending on the level of the course. At the University of Southampton, I have been using the material as follows.

Second Year of MEng/BEng in Electronic Engineering

Chapters 1 and 2 are review material, which the students would be expected to read independently. Lectures then cover the material of Chapters 3 through 7. Some of this material can be considered optional, such as Sections 5.3 and 5.7. Additionally, some constructs could be omitted if time is limited. The single-stuck fault model of Section 11.2 and the principles of test pattern generation in Section 11.3, together with the principles of scan design in Section 12.2, would also be covered in lectures.

Third Year of MEng/BEng in Electronic Engineering

Students would be expected to independently re-read Chapters 4 to 7. Lectures would cover Chapters 8 to 13. Verilog-AMS, Chapter 14, is currently covered in a fourth-year module.

In all years, students need to have access to a SystemVerilog simulator and an RTL synthesis tool in order to use the examples in the text. In the second year, a group design exercise involving synthesis to an FPGA would be an excellent supplement to the material. In the third year at Southampton, all students do an individual project. Some of the individual projects will involve the use of SystemVerilog.

Web Resources

A Web site accompanies *Digital System Design with SystemVerilog* by Mark Zwolinski. Visit the site at zwolinski.org. Here you will find valuable teaching and learning material including all the SystemVerilog examples and links to sites with SystemVerilog tools.

Acknowledgments

I would like to thank all those who pointed out errors in the VHDL versions of this book.

I would also like to thank everyone involved in the commissioning and preparation of this book: Bernard Goodwin and Elizabeth Ryan at Prentice Hall, Madhu Bhardwaj and Ben Kolstad at Glyph International, Susan Fox-Greenberg, who copy edited the text, Danielle Shaw for proof-reading, and Jack Lewis for indexing. Any errors are, however, my fault and not theirs!

Finally, I would like to thank several cohorts of students to whom I have delivered this material and whose comments have encouraged me to think about better ways of explaining these ideas.

About the Author

Mark Zwolinski is a full professor in the School of Electronics and Computer Science, University of Southampton, United Kingdom. He is the author of *Digital System Design with VHDL*, which has been translated into four languages and widely adopted as a textbook in universities worldwide. He has published over 120 refereed papers in technical journals and has been teaching digital design and design automation to undergraduate and graduate students for twenty years.

1 Introduction

In this chapter we will review the design process, with particular emphasis on the design of digital systems using hardware description languages (HDLs) such as SystemVerilog. The technology of CMOS (complementary metal oxide semiconductor) integrated circuits will be briefly revised and programmable logic technologies will be discussed. Finally, the relevant electrical properties of CMOS and programmable logic are reviewed.

1.1 Modern Digital Design

Electronic circuit design has traditionally fallen into two main areas: analog and digital. These subjects are usually taught separately, and electronics engineers tend to specialize in one area. Within these two groupings there are further specializations, such as radio frequency analog design, digital integrated circuit design, and, where the two domains meet, mixed-signal design. In addition, of course, software engineering plays an increasingly important role in embedded systems.

Digital electronics is ever more significant in consumer goods. Cars have sophisticated control systems. Most homes now have personal computers. Products that used to be thought of as analog, such as radio, television, and telephones, are digital. Digital compact discs and MP3s have replaced analog vinyl for recorded audio. With these changes, the lifetimes of products have lessened. In a period of less than a year, new models will probably have replaced all the digital electronic products in your local store.

1.2 Designing with Hardware Description Languages

1.2.1 Design Automation

To keep pace with this rapid change, electronics products have to be designed extremely quickly. Analog design is still a specialized (and well-paid) profession. Digital design has become very dependent on computer-aided design (CAD)—also known as design automation (DA) or electronic design automation (EDA). The EDA tools allow two tasks to be performed: *synthesis*, which is the translation of a specification into an actual implementation of the design; and *simulation* in which the specification or the detailed implementation can be exercised in order to verify correct operation.

Synthesis and simulation EDA tools require that the design be transferred from the designer's imagination into the tools themselves. This can be done by drawing a diagram of the design using a graphical package. This is known as *schematic capture*. Alternatively, the design can be represented in a textual form, much like a software program. Textual descriptions of digital hardware can be written in a modified programming language, such as C, or in HDL. Over the past 30 years, a number of HDLs have been designed. Two HDLs are in common usage today: Verilog and VHDL. Standard HDLs are important because they can be used by different CAD tools from different tool vendors. In the days before Verilog and VHDL, every tool had its own HDL, requiring laborious translation between HDLs, for example, to verify the output from a synthesis tool with another vendor's simulator.

1.2.2 What is SystemVerilog?

SystemVerilog is an HDL. In many respects, an HDL resembles a software programming language, but HDLs have several features not present in languages such as C.

Verilog was first developed in the early 1980s. It is based on Hilo-2, which was a language (and simulator) from Brunel University, UK. The company that first developed Verilog, Gateway Design Automation, was bought out by Cadence. In the early 1990s, Cadence put the language into the public domain, and in 1995, Verilog became an IEEE (Institute of Electrical and Electronics Engineers) standard—1364. In 2001, a new version of the standard was agreed upon, with many additional features, and a further minor revision was agreed upon in 2005. Work is continuing to extend Verilog to system-level modeling. This new language is known as

SystemVerilog, the latest version of which is 3.1a (the number assumes the 1995 version of Verilog was version 1.0 and the 2001 revision was 2.0). The language became an IEEE standard, 1800, in 2005.

Verilog has also been extended to allow modeling of analog circuits (Verilog-A) and again for mixed-signal modeling (Verilog-AMS).

1.2.3 What is VHDL?

During the same period of time, another HDL—VHSIC (Very High Speed Integrated Circuit) HDL or VHDL—was developed for the U.S. Department of Defense and was also standardized by the IEEE as standard 1076. There have been four versions of IEEE 1076, in 1987, 1993, 2002, and 2008. There have been other HDLs, for example, Ella and UDL/I, but now Verilog and VHDL are dominant. Each language has its champions and detractors. Objectively (if it is possible to take a truly unbiased view), both languages have weaknesses, and it is futile getting into arguments about which is best.

1.2.4 Simulation

Another HDL is, however, worthy of note. SystemC uses features of C++ to allow modeling of hardware. At this time it is not possible to predict whether SystemC might supersede SystemVerilog or VHDL. It is, however, worth noting that many of the design style guidelines refer to all three languages.

An HDL has three elements that are seldom present in a programming language: (1) concurrency, (2) representation of time, and (3) representation of structure.

Hardware is intrinsically parallel. Therefore, an HDL must be able to describe actions that happen simultaneously. C (to choose a typical and widely used programming language) is sequential.

Actions in hardware take a finite time to complete. Therefore, mechanisms are needed to describe the passage of time.

The structure of hardware is significant. A C program can consist of functions that are called and, having completed their task, no longer retain any sense of internal state. On the other hand, gates or other hardware structures persist and have a state even when they appear to be doing nothing.

SystemC allows these features to be described in a C-like language. SystemVerilog (and VHDL) have these features built in.

Concurrency, time, and structure lead to another significant difference between an HDL and a programming language. A C program can be compiled and executed on a PC or workstation. SystemVerilog can be compiled, but needs a simulator to execute. The simulator handles the interactions between concurrent elements and models the passage of time. The simulator also keeps track of the state of each structural element. A number of SystemVerilog simulators are available.

Advocates of Verilog often argue that it is an easier language to learn than VHDL. This is debatable (they are probably about the same) for one reason. VHDL has a very well-defined simulation model. Two different VHDL simulators are (almost) guaranteed to produce exactly the same simulations. The SystemVerilog simulation model is more loosely defined. Unless you are very careful, two different SystemVerilog simulators may produce *different* simulations. The intention of this book is to show how to write models of hardware that will simulate and synthesize with predictable behavior. For this reason, this book does not attempt to cover every detail of the SystemVerilog language.

1.2.5 Synthesis

SystemVerilog is a hardware *description* language, not a hardware *design* language. In the 1980s, digital simulation was a mature technology; automatic hardware synthesis was not. (This argument applies equally to VHDL.) It is possible to write models in SystemVerilog that do not and cannot correspond to any physically realizable hardware. Only a subset of the language can be synthesized using current register transfer level (RTL) synthesis tools. Moreover, RTL synthesis tools work by recognizing particular patterns in the code and use those patterns to infer the existence of registers. (RTL synthesis tools also optimize combinational logic, but not much more.) Therefore, the style of SystemVerilog coding is important. An IEEE standard for RTL synthesis—1364.1—was agreed upon in 2002. This defines a subset of Verilog, and the meaning of that subset, of the 2001 revision of Verilog.

The hardware models in this book conform to an application of the 1364.1-2002 RTL synthesis standard to the SystemVerilog language. In other words, we will use the synthesis standard as a style guide.

1.2.6 Reusability

The electronics industry is currently very keen on the idea of reuse. Integrated circuits are so large and complex that it is almost impossible for one team to create a design from scratch. Instead, it is expected that more and more of a design will consist of parts reused from earlier projects or brought in from specialized design companies. Clearly, if a design is to be reused, it has to be versatile. It has to be either so common that everyone will want to use it, or adaptable such that it can be used in a variety of designs.

At a simple level, imagine that you have been asked to design a 4-bit multiplier. This can be done by setting the widths of the inputs and outputs to 4. You would also need to set the widths of some internal registers. At a later date, you might be asked to design a 13-bit adder. At a functional level (or RTL for a synthesizable design), the difference between the two designs is simply a change of input, output, and register widths. Both the new and original designs would have needed simulating and synthesizing. It is possible you might make a mistake in changing your design from 4 to 13 bits. This would require a debugging effort. Imagine instead that you had designed an "n-bit" multiplier. This would be debugged once. When asked to produce the 13-bit multiplier, you would simply plug the parameter "13" into the model and take the rest of the day off! The idea of producing parameterizable designs is therefore very attractive. We will, as far as is possible, design parameterizable, reusable components.

We will also show how to write models that are likely to behave the same way in different simulators and that synthesize with the same results with different synthesis tools. Related to this is a need to ensure that the behavior after synthesis is the same as the behavior before synthesis.

1.2.7 Verification

How do we know that a model accurately describes the hardware that we want to build? Formal verification tools exist, but they are somewhat specialized and difficult to use. Simulation is a much more common technique. In simulation, we try to give a design as wide a range of inputs as possible in order to cover everything that could happen to that design. This approach can apply to each individual part and to the system as a whole. As the hardware model gets larger, the range of possible behaviors increases. Therefore, it become harder to exhaustively test the model and the simulation time grows. This is a disadvantage of using simulation as a verification tool.

In this book, a number of examples are given. You are encouraged to investigate these models by running simulations. To do this, you will need to provide test stimuli. One of the factors that has made SystemVerilog so important is the ability to use the language itself to describe these test stimuli. This may seem obvious—in the 1980s, this was an innovation; hardware modeling and test definitions were usually done using entirely different languages. Later, hardware *verification* languages such as Vera and *e* were developed, but many of their features have been absorbed into SystemVerilog.

In the jargon, the test stimuli are defined in a "testbench." A testbench is a piece of SystemVerilog code that (conceptually) surrounds the model and defines the universe as seen by that model. Therefore, a testbench has no inputs or outputs (and can be identified by this feature). Within a testbench, you might write a description of a clock generator and define a sequence of inputs. You might also check the responses from the model. Within the simulator, you can display the waveforms of signals, even signals deep within the design.

Writing testbenches requires a coding style different than hardware modeling. A testbench does not represent a piece of real hardware. Indeed, you should never attempt to synthesize a testbench. You will just get pages of warning messages. Again, the SystemVerilog simulation problem arises here. A testbench may behave differently for different simulators. We will try to minimize this problem, but it is a less precise art than writing portable RTL models.

Simulation can help to ensure that your design implements the specification as accurately as is humanly possible (and humans are never capable of perfection). We can, with a bit of luck, assume that the synthesis process correctly translates a SystemVerilog description into gates and flip-flops. When thousands or millions of the final integrated circuit are manufactured, it is inevitable that defects will occur in a small (we hope) number of chips. These defects can be caused by, for example, dirt or imperfections in the silicon. If these defects cause the electrical behavior of the circuit to change, the circuit will not work correctly. Such faulty circuits need to be detected at the time of manufacture. *Fault simulation* allows potential faults in a circuit to be modeled and rapidly simulated. Another testbench or set of testbenches and a fault simulator are needed to determine a minimal set of test vectors to uncover all possible faults.

1.2.8 Design Flow

Most digital systems are sequential, that is, they have states, and the outputs depend on the present state. Some early designs of computers were asynchronous; in other words, the transition to a new state happened as soon as inputs had stabilized. For many years, digital systems have tended to be synchronous. In a synchronous system, the change of state is triggered by one or more clock signals. In order to design reliable systems, formal design methodologies have been defined. The design of a (synchronous sequential) digital system using discrete gates would therefore proceed as follows.

1. Write a specification.
2. If necessary, partition the design into smaller parts and write a specification for each part.

3. From the specification, draw a state machine chart. This shows each state of the system and the input conditions that cause a change of state, together with the outputs in each state.
4. Minimize the number of states. This is optional and may not be useful in all cases.
5. Assign Boolean variables to represent each state.
6. Derive the next state and output logic.
7. Optimize the next state and output logic to minimize the number of gates needed.
8. Choose a suitable placement for the gates in terms of which gates share integrated circuits and where each integrated circuit is placed on the printed circuit board.
9. Design the routing between the integrated circuits.

In general, steps 1 and 2 cannot be avoided. This is where the creativity of the designer is needed. Most books on digital design concentrate on steps 3 to 7. Steps 8 and 9 can be performed manually, but placement and routing was one of the first tasks to be successfully automated. It is possible to simulate the design at different stages if it is converted into a computer-readable form. Typically, in order to perform the placement and routing, a schematic capture program would be used at around step 7, such that the gate-level structure of the circuit would be entered. This schematic could be converted to a form suitable for a logic simulator. After step 9 has been completed, the structure of the circuit, including any delays generated by the resistance and capacitance of the interconnect, could be extracted and again simulated.

The implementation of digital designs on ASICs or field programmable gate areas (FPGAs) therefore involves the configuration of connections between predefined logic blocks. As noted, we cannot avoid steps 1 and 2, and steps 8 and 9 can be done automatically. The use of an HDL, here SystemVerilog, means that the design can be entered into a CAD system and simulated at step 3 or 4, rather than step 7. So-called RTL synthesis tools automate steps 6 and 7. Step 5 can be automated, but now the consequences of a particular state assignment can be assessed very quickly. Behavioral synthesis tools are starting to appear that automate the process from about step 2 onwards. Figure 1.1 shows the overall design flow for RTL synthesis-based design.

Because of this use of EDA tools to design ASICs and FPGAs, a book such as this can concentrate on higher-level aspects of design, in particular the description

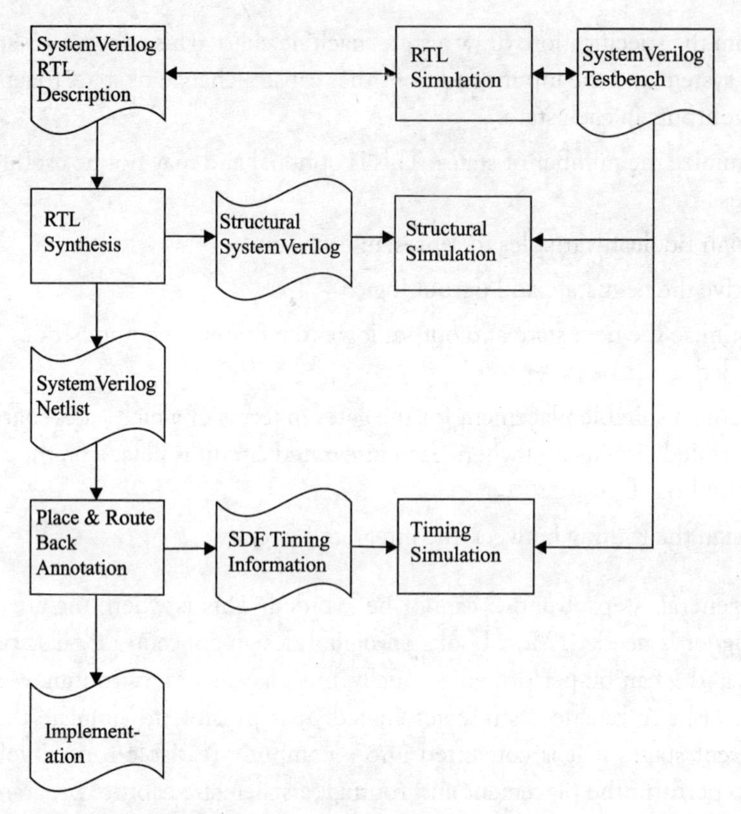

Figure 1.1 RTL synthesis design flow.

of functional blocks in an HDL. Many books on digital design describe multiple output and multi-level logic minimization, including techniques such as the Quine–McCluskey algorithm. Here, we assume that a designer may occasionally wish to minimize expressions with a few variables and a single output, but if a complex piece of combinational logic is to be designed, a suitable EDA tool is available that will perform the task quickly and reliably.

1.3 CMOS Technology

1.3.1 Logic Gates

The basic building blocks of digital circuits are *gates*. A gate is an electronic component with a number of inputs and, generally, a single output. The inputs and the outputs are normally in one of two states: logic 0 or logic 1. These logic values are

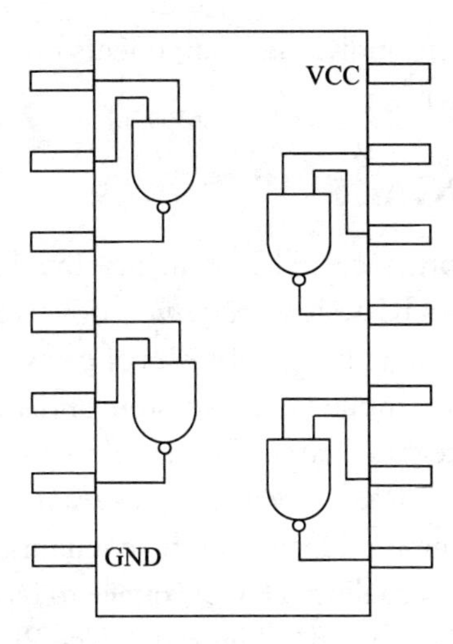

Figure 1.2 Small-scale integrated circuit.

represented by voltages (for instance, 0 V for logic 0 and 2.5 V for logic 1) or currents. The gate itself performs a logical operation using all of its inputs to generate the output. Ultimately, of course, digital gates are really analog components, but for simplicity we tend to ignore their analog nature.

It is possible to buy a single integrated circuit containing, say, four identical gates, as shown in Figure 1.2. (Note that two of the connections are for the positive and negative power supplies to the device. These connections are not normally shown in logic diagrams.) A digital system could be built by connecting hundreds of such devices together—indeed many systems have been designed in that way. Although the individual integrated circuits might cost as little as 10 cents each, the cost of designing the printed circuit board for such a system and the cost of assembling the board are very significant, and this design style is no longer cost effective.

Much more complicated functions are available as mass-produced integrated circuits, ranging from flip-flops to microprocessors. With increasing complexity comes flexibility—a microprocessor can be programmed to perform a near-infinite variety of tasks. Digital system design therefore consists of, in part, of taking standard components and connecting them together. Inevitably, however, some aspect of the functionality will not be available as a standard device. The designer is then left with the choice of implementing this functionality from discrete gates or designing a specialized integrated circuit to perform that task. While this latter task may appear daunting, it should be remembered that the cost of a system will depend to a great extent not on the cost of the individual components but on the cost of connecting those components together.

1.3.2 ASICs and FPGAs

The design of a high-performance, full-custom integrated circuit (IC) is, of course, a difficult task. In full-custom IC design, *everything*, down to and including individual transistors, may be designed (although libraries of parts are, of course, used).

The term ASIC is often applied to full-custom and semi-custom integrated circuits. Another class of integrated circuit is that of *programmable logic*. The earliest programmable logic devices (PLDs) were *programmable logic arrays* (PLAs). These consist of arrays of uncommitted logic and the configuration of the array is determined by applying a large (usually negative) voltage to individual connections. The general structure of a PLA is shown in Figure 1.3. The PLA has a number of inputs (A, B, C) and outputs (X, Y, Z), an AND-plane, and an OR-plane. Connections between the inputs and the product terms (P, Q, R, S) and between the product terms and outputs are shown; the remaining connections have been removed as part of the programming procedure. Some PLAs may be reprogrammed electrically, or by restoring the connections by exposing the device to ultraviolet light. Programmable array logic (PAL) extends the idea of PLAs to include up to 12 flip-flops. In recent years, programmable devices have become much more complex and include complex PLDs (CPLDs) and FPGAs. FPGAs are described in more detail in Section 1.4.

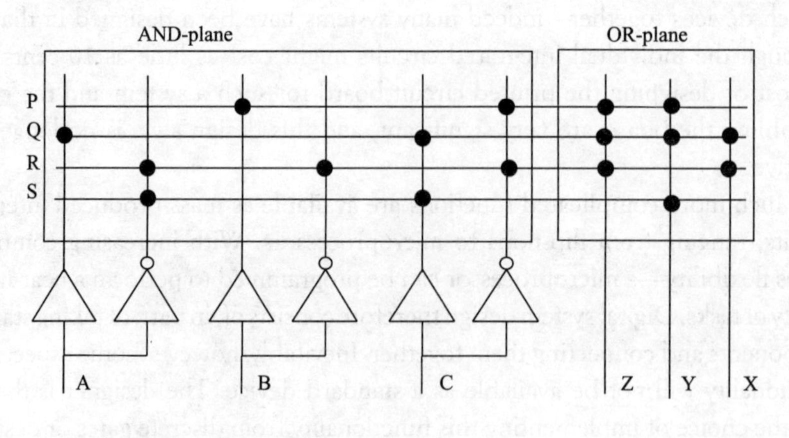

Figure 1.3 PLA structure.

Even digital gates can be thought of as analog circuits. The design of individual gates is therefore a circuit design problem. Hence, there exists a wide variety of possible circuit structures. Very early digital computers were built using vacuum tubes. These gave way to transistor circuits in the 1960s and 1970s. There are two major types of transistor: bipolar junction transistors (BJTs) and field effect transistors (FETs). Logic families such as transistor–transistor logic (TTL) and emitter–collector logic (ECL) use BJTs. Today, the dominant (but not exclusive) technology is CMOS, which uses FETs. CMOS derives its name from the particular type of FET used—the MOSFET (metal oxide semiconductor FET). CMOS therefore stands for complementary MOS, as two types of MOS device are used. MOS is, in fact, a misnomer; a better term is IGFET (insulated gate FET).

The structure of an n-type MOS (NMOS) transistor is shown in Figure 1.4, which is not drawn to scale. The substrate is the silicon wafer that has been doped to make it p-type. The thickness of the substrate is therefore significantly greater than the other transistor dimensions. Two heavily doped regions of n-type silicon are created for each transistor. These form the source and drain. In fact, the *source* and *drain* are interchangeable, but by convention the drain–source voltage is usually positive. Metal connections are made to the source and drain. The polycrystalline silicon (polysilicon) gate is separated from the rest of the device by a layer of silicon dioxide insulator. Originally, the gate would have been metal; hence, the name MOS was derived from the structure of the device (metal oxide semiconductor).

When the gate voltage is the same as the source voltage, the drain is insulated from the source. As the gate voltage rises, the gate–oxide–semiconductor sandwich acts as a capacitor, and negative charge builds up on the surface of the semiconductor. At a critical *threshold voltage* the charge is sufficient to create a channel of n-type silicon between the source and the drain. This acts as a conductor between the source and the drain. Therefore, the NMOS transistor can be used as a

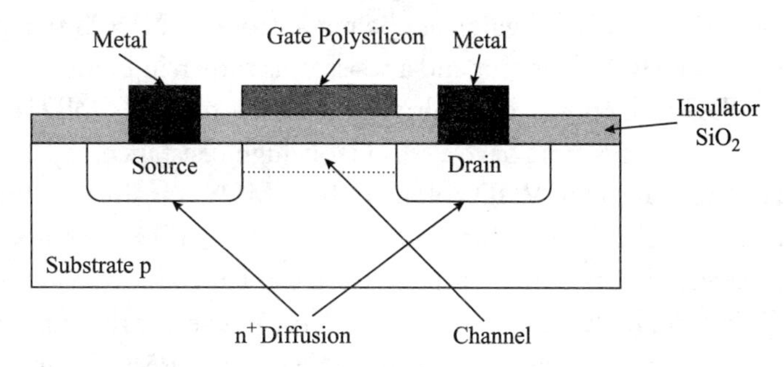

Figure 1.4 NMOS transistor structure.

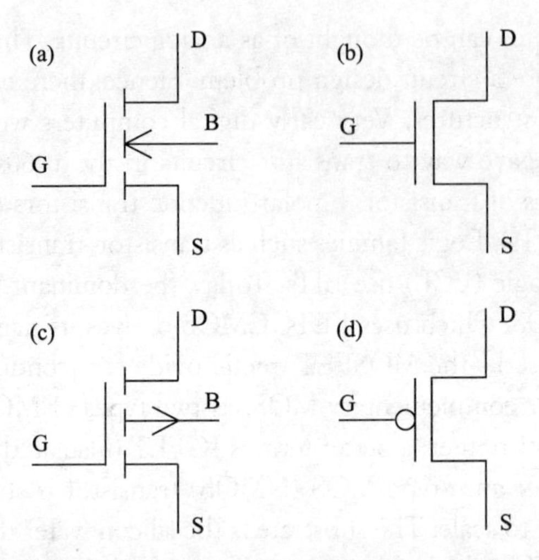

Figure 1.5 MOS transistor symbols: (a) and (b) NMOS, (c) and (d) PMOS.

switch that is open when the gate voltage is low and closed when the gate voltage is high.

A p-type MOS (PMOS) transistor is formed by creating heavily doped p-type drain and source regions in an n-type substrate. A PMOS transistor conducts when the gate voltage is low and does not conduct when the gate voltage is high.

Symbols for NMOS transistors are shown in Figure 1.5(a) and (b). The substrate is also known as the *bulk*, hence the symbol B. In digital circuits, the substrate of NMOS transistors is always connected to ground (logic 0) and hence can be omitted from the symbol, as shown in Figure 1.5(b). Symbols for PMOS transistors are shown in Figure 1.5(c) and (d). Again the bulk connection is not shown in Figure 1.5(d) because in digital circuits the substrate of a PMOS transistor is always connected to the positive supply voltage (logic 1).

A logical inverter (a NOT gate) can be made from an NMOS transistor and a resistor, or from a PMOS transistor and a resistor, as shown in Figure 1.6(a) and (b), respectively. VDD is the positive supply voltage (3.3 V to 5 V); GND is the ground connection (0 V). The resistors have a reasonably high resistance, say, 10 kΩ. When IN is at logic 1 (equal to the VDD voltage), the NMOS transistor in Figure 1.6(a) acts as a closed switch. Because the resistance of the NMOS transistor, when it is conducting, is much less than that of the resistor, OUT is connected to GND, giving a logic 0 at that node. In the circuit of Figure 1.6(b), a logic 1 at IN causes the PMOS transistor to act as an open switch. The resistance of the PMOS transistor is now

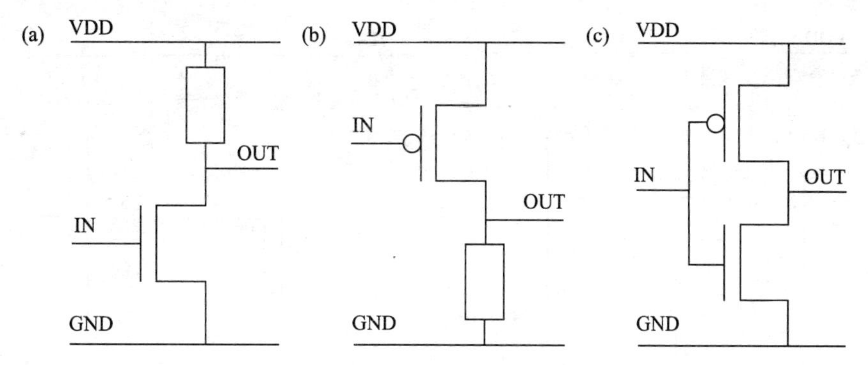

Figure 1.6 MOS inverters: (a) NMOS, (b) PMOS, (c) CMOS.

much greater than that of the resistor, so OUT is connected to GND via the resistor. Again, a logic 0 is asserted at OUT.

A logic 0 at IN causes the opposite effect. The NMOS transistor becomes an open switch, causing OUT to be connected to VDD by the resistor; the PMOS transistor becomes a closed switch with a lower resistance than the resistor and again OUT is connected to VDD.

Figure 1.6(c) shows a CMOS inverter. Here, both PMOS and NMOS transistors are used. A logic 1 at IN will cause the NMOS transistor to act as a closed switch and the PMOS transistor to act as an open switch, giving a 0 at OUT. A logic 0 will have the opposite effect: the NMOS transistor will be open and the PMOS transistor will be closed. The name CMOS comes from complementary MOS—the NMOS and PMOS transistors complement each other.

Current flows in a semiconductor as electrons move through the crystal matrix. In p-type semiconductors, it is convenient to think of the charge being carried by the absence of an electron, a "hole." The mobility of holes is less than that of electrons (i.e., holes move more slowly through the crystal matrix than do electrons). The effect of this is that the gain of a PMOS transistor is less than that of a same-sized NMOS transistor. Thus, to build a CMOS inverter with symmetrical characteristics, in the sense that a 0 to 1 transition happens at the same rate as a 1 to 0 transition, requires that the gain of the PMOS and NMOS transistors be made the same. This is done by varying the widths of the transistors (assuming the lengths are the same) such that the PMOS transistor is about 2.5 times as wide as the NMOS transistor. As will be seen, this effect is compensated for in CMOS NAND gates, where similarly sized NMOS and PMOS transistors can be used. CMOS NOR gates, however,

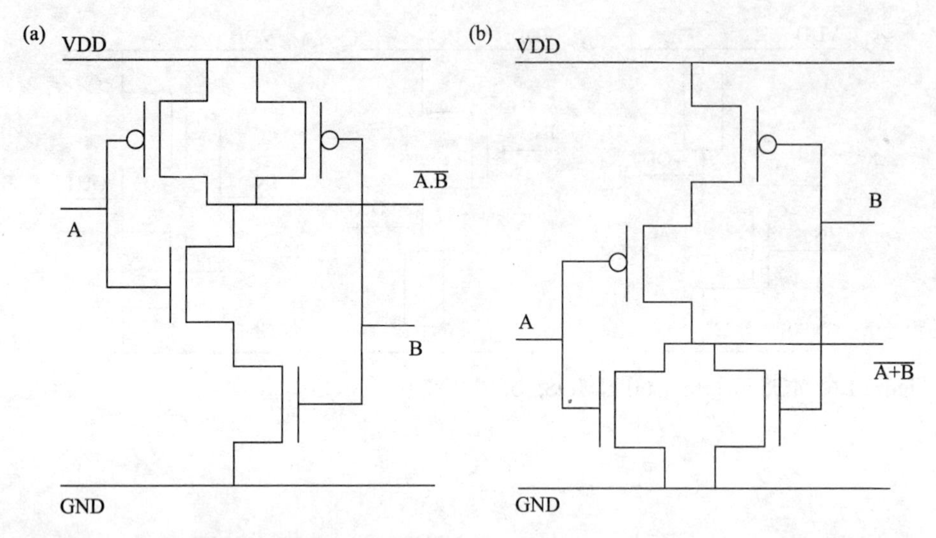

Figure 1.7 (a) CMOS NAND; (b) CMOS NOR.

do require the PMOS transistors to be scaled. Hence, NAND gate logic is often preferred for CMOS design.

Two-input CMOS NAND and NOR gates are shown in Figure 1.7(a) and (b), respectively. The same reasoning as used in the description of the inverter may be applied. A logic 1 causes an NMOS transistor to conduct and a PMOS transistor to be open; a logic 0 causes the opposite effect. NAND and NOR gates with three or more inputs can be constructed using similar structures. Note that in a NAND gate all the PMOS transistors must have a logic 0 at their gates for the output to go high. As the transistors are working in parallel, the effect of the lower mobility of holes on the gain of the transistors is overcome.

Figure 1.8 shows a CMOS AND–OR–Invert structure. The function $\overline{A.B + C.D}$ can be implemented using 8 transistors compared with the 14 needed for three NAND/NOR gates and an inverter.

A somewhat different type of structure is shown in Figure 1.9(a). This circuit is a three-state buffer. When the *EN* input is at logic 1, and the $\overline{EN}$ input is at logic 0, the two inner transistors are conducting, and the gate inverts the *IN* input as normal. When the *EN* input is at logic 0 and the $\overline{EN}$ input is at logic 1, neither of the two inner transistors is conducting, and the output floats. The $\overline{EN}$ input is derived from *EN* using a standard CMOS inverter. An alternative implementation of a three-state buffer is shown in Figure 1.9(b). Here a transmission gate follows the CMOS inverter. The NMOS and PMOS transistors of the transmission gate are

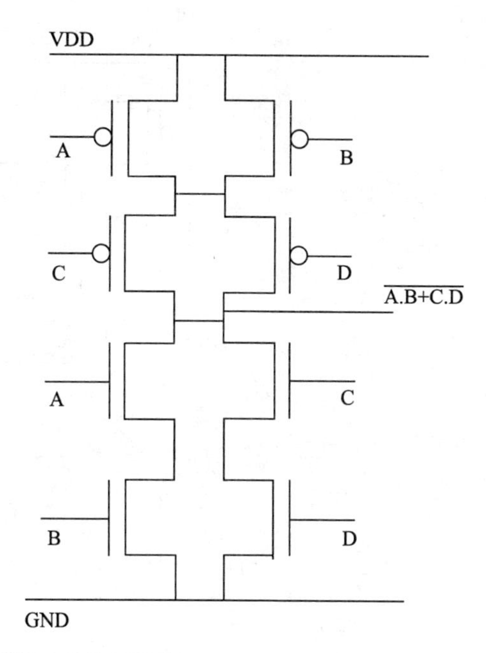

Figure 1.8 CMOS AND-OR-INVERT.

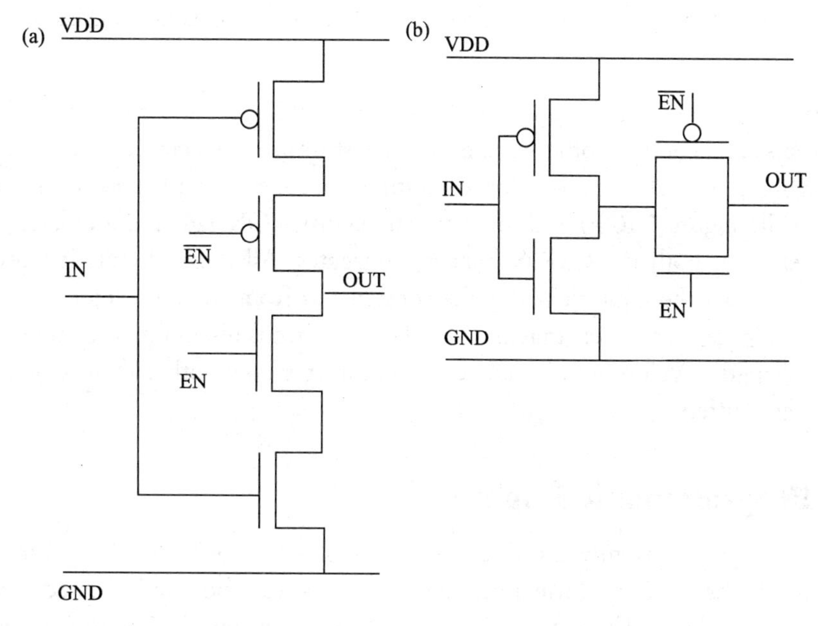

Figure 1.9 CMOS three-state buffer.

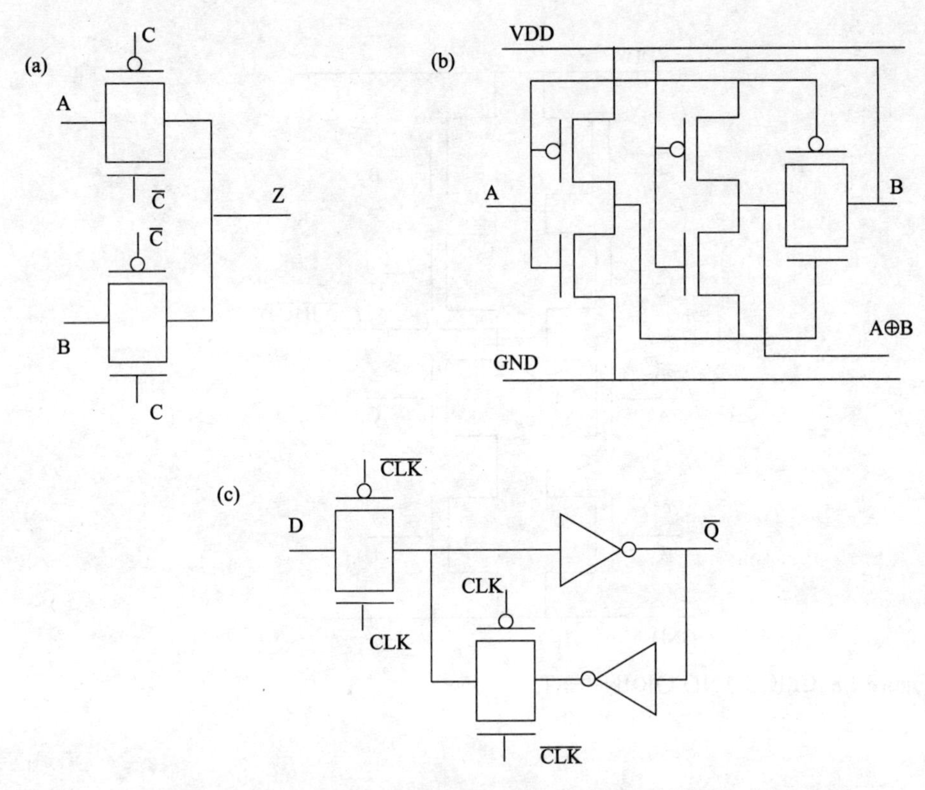

Figure 1.10 CMOS transmission gate circuits. (a) Multiplexer; (b) XOR; (c) D latch.

controlled by complementary signals. When EN is at logic 1 and $\overline{EN}$ is at logic 0, both transistors conduct; otherwise, both transistors are open circuit.

Figure 1.10(a) shows a two-input multiplexer constructed from transmission gates, while Figure 1.10(b) and (c) show an exclusive OR gate and a D latch, respectively, that both use CMOS transmission gates. All these circuits use fewer transistors than the equivalent circuits constructed from standard logic gates. It should be noted, however, that the simulation of transmission gate circuits can be problematic. We do not give any examples in this book, other than of general three-state buffers.

1.4 Programmable Logic

While CMOS is currently the dominant technology for integrated circuits, for reasons of cost and performance, many designs can be implemented using programmable logic. The major advantage of *programmable logic* is the speed of implementation. A PLD can be configured on a desktop in seconds, or at most minutes. The fabrication of an integrated circuit can take several weeks. The cost per device of a circuit built in programmable logic may be greater than that of a custom IC, and the performance, in terms of both speed and functionality, is likely to be less impressive than that of CMOS. These apparent disadvantages are often outweighed by the ability to rapidly produce working ICs. Thus, programmable logic is suited to prototypes, but also increasingly to small production volumes.

One recent application of programmable devices is as *reconfigurable logic.* A system may perform different functions at different points in time. Instead of having all the functionality available all the time, one piece of hardware may be reconfigured to implement the different functions. New functions, or perhaps better versions of existing functions, could be downloaded from the Internet. Such applications are likely to become more common in the future.

There are a number of different technologies used for programmable logic by different manufacturers. The simplest devices, PLAs, consist of two programmable planes, as shown in Figure 1.3. In reality, both planes implement a NOR function. The device is programmed by breaking connections. Most simple programmable devices use some form of floating gate technology. Each connection in the programmable planes consists of a MOS transistor. This transistor has two gates—one is connected to the input, while the second, between the first gate and the channel, floats. When the appropriate negative voltage is applied to the device, the floating gate can have a large charge induced on it. This charge will exist indefinitely. If the charge exists on the floating gate, the device is disabled; if the charge is not there, the device acts as a normal transistor. The mechanisms for putting the charge on the device include *avalanche* or *hot electron injection* (EPROM) and *Fowler–Nordheim tunneling* (EEPROM and Flash devices). These devices can be reprogrammed electrically.

PALs have a programmable AND plane and a fixed OR plane, and usually include registers, as shown in Figure 1.11. CPLDs effectively consist of a number of PAL-like macrocells that can communicate through programmable interconnect, as shown in Figure 1.12

More complex still are FPGAs. Xilinx FPGAs are implemented in static RAM technology. Unlike other programmable logic, the configuration is therefore volatile and must be restored each time power is applied to the circuit. Again, these FPGAs consist of arrays of logic cells. One such cell is shown in Figure 1.13 Each of these cells can be programmed to implement a range of combinational and sequential functions. In addition to these logic cells, there exists programmable interconnect, including three-state buffers.

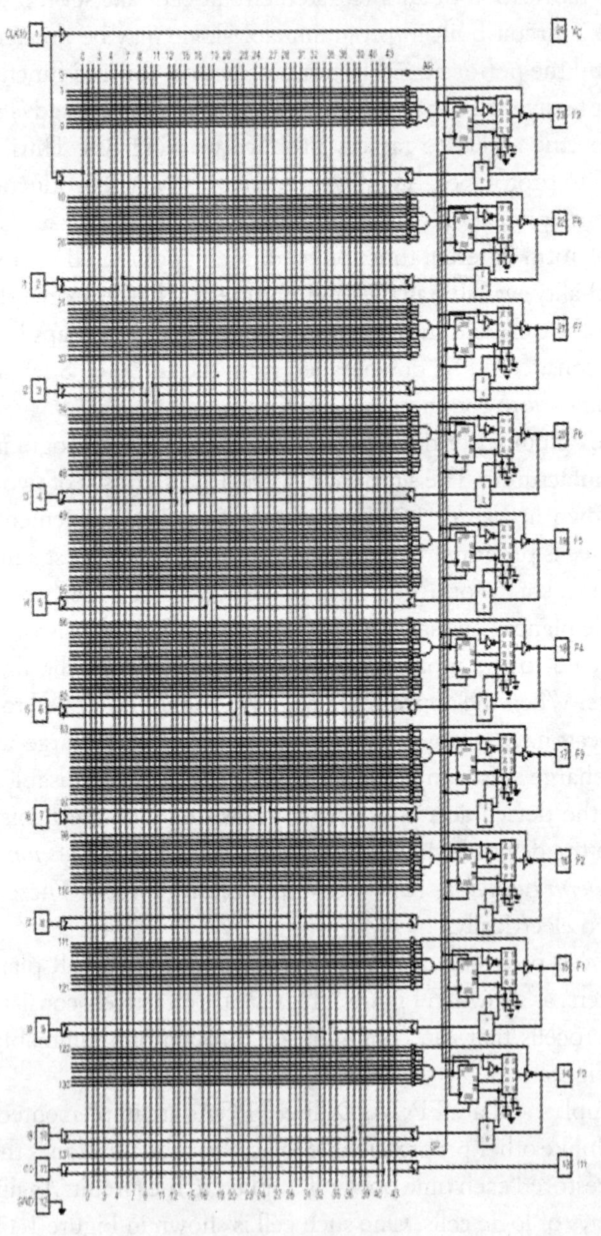

Figure 1.11 PAL structure (Copyright © Lattice Semiconductor Corporation. Reprinted with permission of the copyright owner. All other rights reserved.)

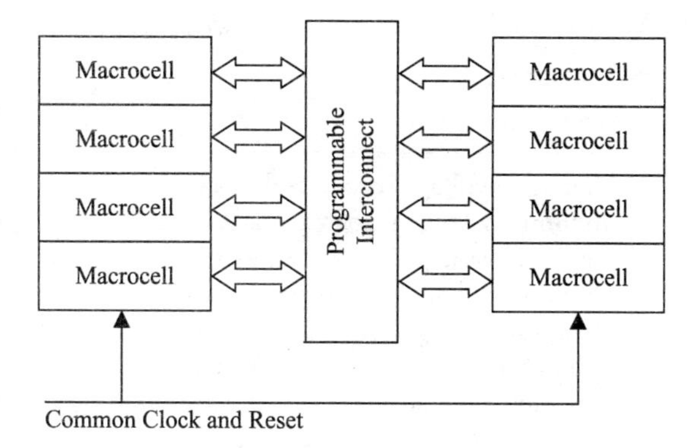

Figure 1.12 CPLD structure.

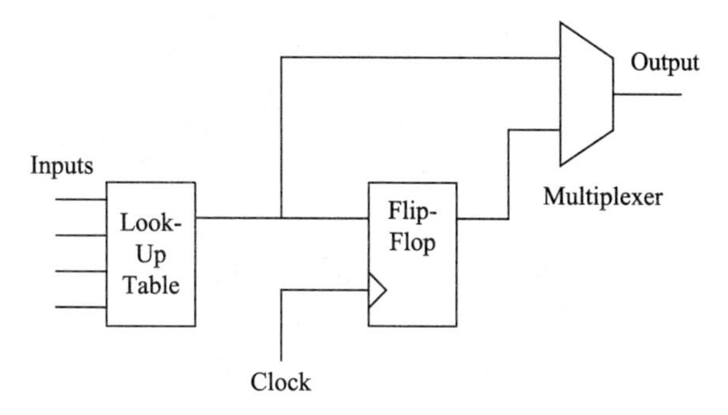

Figure 1.13 FPGA logic cell.

1.5 Electrical Properties

1.5.1 Noise Margins

Although it is common to speak of a logic 1 being, say, 2.5 V and a logic 0 being 0 V, in practice a range of voltages represents a logic state. A range of voltages may be recognized as a logic 1, and similarly one voltage from a particular range may be generated for a logic 1. Thus, we can describe the logic states in terms of the voltages shown in Table 1.1.

The transfer characteristic for a CMOS inverter is illustrated in Figure 1.14. The *noise margin* specifies how much noise, from electrical interference, can be added to a signal before a logic value is misinterpreted. From Table 1.1 it can be seen that the maximum voltage that a gate will generate to represent a logic 0 is 0.75 V. Any voltage up to 1.05 V is, however, recognized as a logic 0. Therefore, there is a "spare" 0.3 V, and any noise added to a logic 0 within this band will be accepted. Similarly, the difference between the minimum logic 1 voltage generated

Table 1.1 Typical Voltage Levels for CMOS Circuits with a Supply Voltage of 2.5 V

Parameter	Description	Typical CMOS Value
$V_{IH\,\max}$	Maximum voltage recognized as a logic 1	2.5 V
$V_{IH\,\min}$	Minimum voltage recognized as a logic 1	1.35 V
$V_{IL\,\max}$	Maximum voltage recognized as a logic 0	1.05 V
$V_{IL\,\min}$	Minimum voltage recognized as a logic 0	0.0 V
$V_{OH\,\max}$	Maximum voltage generated as a logic 1	2.5 V
$V_{OH\,\min}$	Minimum voltage generated as a logic 1	1.75 V
$V_{OL\,\max}$	Maximum voltage generated as a logic 0	0.75 V
$V_{OL\,\min}$	Minimum voltage generated as a logic 0	0.0 V

and the minimum recognized is 0.4 V. The noise margins are calculated as:

$$NM_L = V_{IL\,\max} - V_{OL\,\max}$$

$$NM_H = V_{OH\,\min} - V_{IH\,\min}$$

In general, the bigger the noise margin, the better.

1.5.2 Fan-Out

The fan-out of a gate is the number of other gates that it can drive. Depending on the technology, there are two ways to calculate the fan-out. If the input to a gate is resistive, as is the case with TTL or anti-fuse technology, the fan-out is calculated as the ratio of the current that a gate can output to the amount of current required

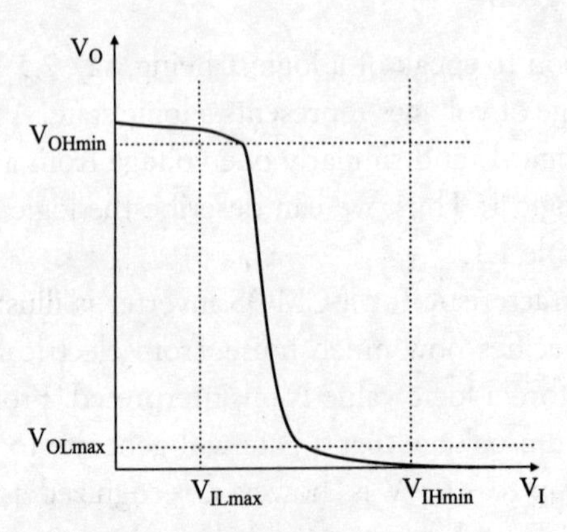

Figure 1.14 Transfer characteristic of a CMOS inverter.

Table 1.2 Input and Output Currents for 74ALS Series TTL Gates

$I_{IH\max}$	Maximum logic 1 input current	20 μA
$I_{IL\max}$	Maximum logic 0 input current	-100μA
$I_{OH\max}$	Maximum logic 1 output current	-400μA
$I_{OL\max}$	Maximum logic 0 output current	8 mA

to switch the input of a gate. For example, 74ALS series gates have the input and output currents specified in Table 1.2.

Two fan-out figures can be calculated:

$$Logic\ 1\ fan\text{-}out = \frac{I_{OH\max}}{I_{IH\max}} = \frac{400\mu A}{20\mu A} = 20$$

$$Logic\ 0\ fan\text{-}out = \frac{I_{OL\max}}{I_{IL\max}} = \frac{8mA}{100\mu A} = 80$$

Obviously, the smaller of the two figures must be used.

CMOS gates draw almost no DC input current because there is no DC path between the gate of a transistor and the drain, source, or substrate of the transistor. Therefore, it would appear that the fan-out of CMOS circuits is very large. A different effect applies in this case. Because the gate and substrate of a CMOS gate form a capacitor, it takes a finite time to charge that capacitor, and hence the fan-out is determined by how fast the circuit is required to switch. In addition, the interconnect between two gates has a capacitance. In high-performance circuits, the effect of the interconnect can dominate that of the gates themselves. Obviously, the interconnect characteristics cannot be estimated until the final layout of the circuit has been completed.

Figure 1.15(a) shows one CMOS inverter driving another. Figure 1.15(b) shows the equivalent circuit. If the first inverter switches from a logic 1 to a logic 0 at $t = 0$, and if we assume that the resistance of the NMOS transistor is significantly less than the resistance of the PMOS transistor, V_O is given by:

$$V_O = V_{DD}e^{-t/R_N C_G}.$$

From Table 1.1, the minimum value of V_O that would be recognized as a logic 1 is 1.35 V, and the maximum value of V_O that would be recognized as a logic 0 is 1.05 V. For example, if V_{DD} is 2.5 V, R_N is 100 Ω, and C_G is 100 pF, we can see that

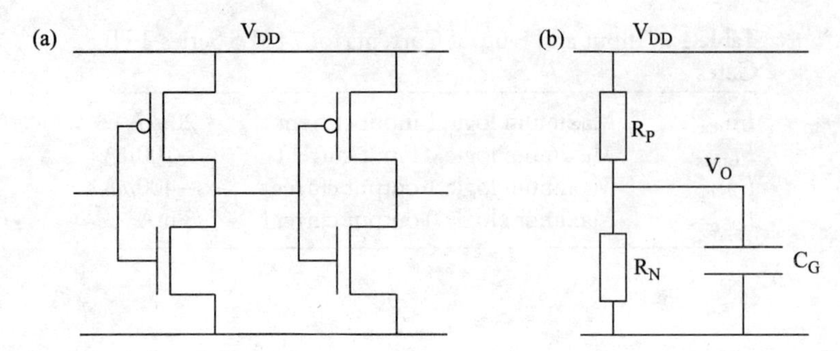

Figure 1.15 (a) CMOS inverter driving CMOS inverter; (b) equivalent circuit.

the time taken for V_O to drop from 1.35 V to 1.05 V is given by:

$$t = -100 \times 100 \times 10^{-12} \times \ln \frac{1.05}{2.5} + 100 \times 100 \times 10^{-12} \times \ln \frac{1.35}{2.5}$$

$$= 2.5 \text{ ns}$$

If two inverters are driven, the capacitive load doubles, so the switching time doubles. Therefore, although a CMOS gate can drive an almost unlimited number of other gates at a fixed logic level, the fan-out is limited by the speed required of the circuit.

Summary

Digital design is no longer a matter of taking small-scale ICs and connecting them together. PLDs are an important alternative to full-custom ICs. A number of different technologies exist for PLDs. These different technologies impose different constraints on the designer.

Further Reading

The best source of information about different families of programmable logic is the manufacturers themselves. Entire data books are now available on the Web. These generally include electrical information, design advice, and hints for programming using SystemVerilog or Verilog. In general, it is easy to guess the Web addresses; for example, Xilinx is at xilinx.com and Actel is at actel.com.

Exercises

1.1 Find examples of the following components in a 74ALS/74AHC data book (or on the Web):

- 4-bit universal shift register
- 4-bit binary counter
- 8-bit priority encoder
- 4-bit binary adder
- 4-bit ALU (Arithmetic and Logic Unit)

1.1 Find examples of PLDs, CPLDs, and FPGAs from manufacturers' data books or from the Web. Compare the following factors:

- Technologies
- Performance
- Cost
- Programmability (e.g., use of SystemVerilog)
- Testability

1.2 How is SystemVerilog used in the design process?

1.3 FPGAs are available in a number of sizes. Given that smaller FPGAs will be cheaper, what criteria would you use to estimate the required size of an FPGA, prior to detailed design?

1.4 A digital system may be implemented in a number of different technologies. List the main types available and comment on the advantages and disadvantages of each option. If you were asked to design a system with about 5,000 gates and which was expected to sell about 10,000 units, which hardware option would you choose and why?

2 Combinational Logic Design

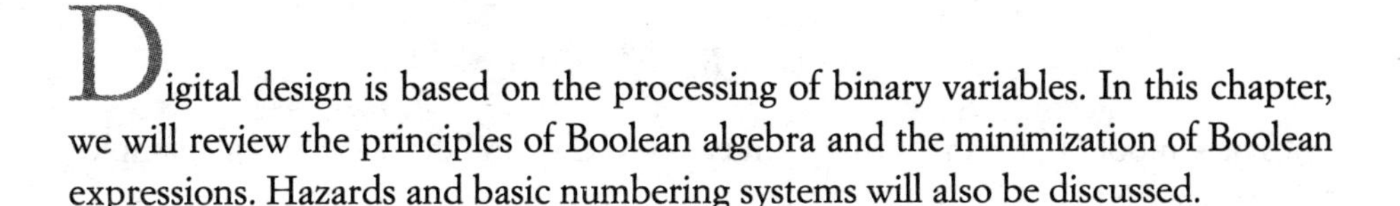

Digital design is based on the processing of binary variables. In this chapter, we will review the principles of Boolean algebra and the minimization of Boolean expressions. Hazards and basic numbering systems will also be discussed.

2.1 Boolean Algebra

2.1.1 Values

Digital design uses a two-value algebra. Variables can take one of two values that can be represented by

ON and OFF,

TRUE and FALSE,

1 and 0.

2.1.2 Operators

The algebra of two values, known as Boolean algebra, after George Boole (1815–1864), has five basic operators. In decreasing order of precedence (i.e., in the absence of parentheses, operations at the top of the list should be evaluated first) these are:

1. NOT
2. AND
3. OR
4. IMPLIES
5. EQUIVALENCE

The last two operators are not widely used in digital design. These operators can be used to form expressions. For example:

$$A = 1$$

$$B = C \text{ AND } 0$$

$$F = \overline{(A + B \cdot C)}$$

$$Z = (\bar{A} + B) \cdot (A + \bar{B})$$

The symbol "+" means "OR," "." means "AND," and the overbar, for example, "$\bar{A}$," means "NOT A."

2.1.3 Truth Tables

The meaning of an operator or expression can be described by listing all the possible values of the variables in that expression, together with the value of the expression in a *truth table*. The truth tables for the three basic operators are given in Tables 2.1, 2.2, and 2.3.

In digital design, three further operators are commonly used: NAND (Not AND), NOR (Not OR), and XOR (eXclusive OR); see Tables 2.4, 2.5, and 2.6.

The XNOR ($\overline{A \oplus B}$) operator is also used occasionally. XNOR is the same as EQUIVALENCE.

Table 2.1 NOT Operation

A	NOT A ($\bar{A}$)
0	1
1	0

Table 2.2 AND Operation

A	B	A AND B $(A \cdot B)$
0	0	0
0	1	0
1	0	0
1	1	1

Table 2.3 OR Operation

A	B	A OR B $(A + B)$
0	0	0
0	1	1
1	0	1
1	1	1

Table 2.4 NAND Operation

A	B	A NAND B $\overline{(A \cdot B)}$
0	0	1
0	1	1
1	0	1
1	1	0

Table 2.5 NOR Operation

A	B	A NOR B $\overline{(A + B)}$
0	0	1
0	1	0
1	0	0
1	1	0

Table 2.6 XOR Operation

A	B	A XOR B $(A \oplus B)$
0	0	0
0	1	1
1	0	1
1	1	0

2.1.4 Rules of Boolean Algebra

There are a number of basic rules of Boolean algebra that follow from the precedence of the operators.

1. Commutativity

$$A + B = B + A$$

$$A \cdot B = B \cdot A$$

2. Associativity

$$A + (B + C) = (A + B) + C$$

$$A \cdot (B \cdot C) = (A \cdot B) \cdot C$$

3. Distributivity

$$A \cdot (B + C) = A \cdot B + A \cdot C$$

In addition, some basic relationships can be observed from the previous truth tables.

$$\bar{\bar{A}} = A$$

$$A \cdot 1 = A \qquad A + 0 = A$$

$$A \cdot 0 = 0 \qquad A + 1 = 1$$

$$A \cdot A = A \qquad A + A = A$$

$$A \cdot \bar{A} = 0 \qquad A + \bar{A} = 1$$

The right-hand column can be derived from the left-hand column by applying the *principle of duality*. The principle of duality states that if each AND is changed to an OR, each OR to an AND, each 1 to 0, and each 0 to 1, the value of the expression remains the same.

2.1.5 De Morgan's Law

There is a very important relationship that can be used to rewrite Boolean expressions in terms of NAND or NOR operations: de Morgan's Law. This is expressed as

$$\overline{(A \cdot B)} = \bar{A} + \bar{B} \quad \text{or} \quad \overline{(A + B)} = \bar{A} \cdot \bar{B}$$

2.1.6 Shannon's Expansion Theorem

Shannon's expansion theorem can be used to manipulate Boolean expressions.

$$
\begin{aligned}
F(A, B, C, D, \ldots) &= A \cdot F(1, B, C, D, \ldots) + \bar{A} \cdot F(0, \mathrm{B}, \mathrm{C}, \mathrm{D}, \ldots) \\
&= (A + F(0, B, C, D, \ldots)) \cdot (\bar{A} + F(1, B, C, D, \ldots))
\end{aligned}
$$

$F(1, B, C, D, \ldots)$ means that all instances of A in F are replaced by a logic 1.

2.2 Logic Gates

The basic symbols for one and two input logic gates are shown in Figure 2.1. Three and more inputs are shown by adding extra inputs (but note that there is no such thing as a three input XOR gate). The ANSI/IEEE symbols can be used instead

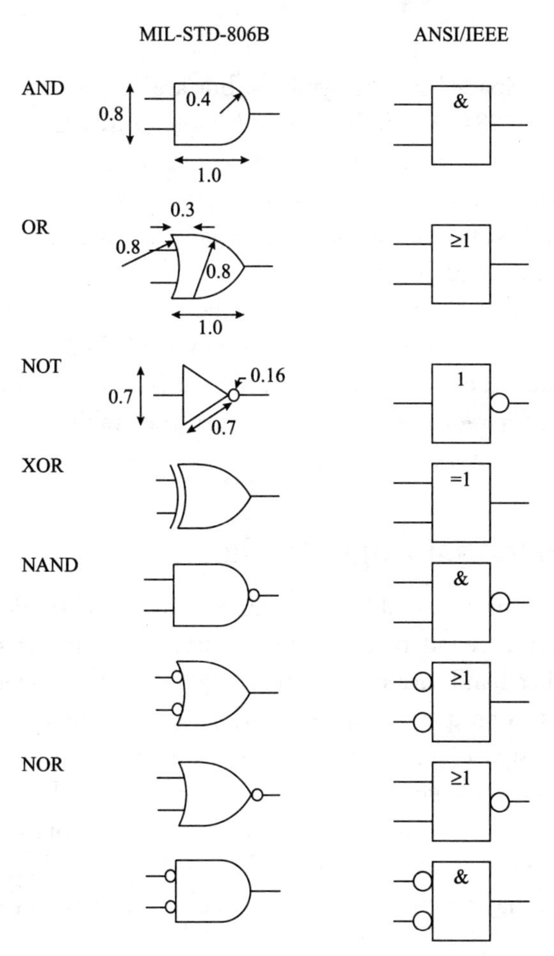

Figure 2.1 Logic symbols.

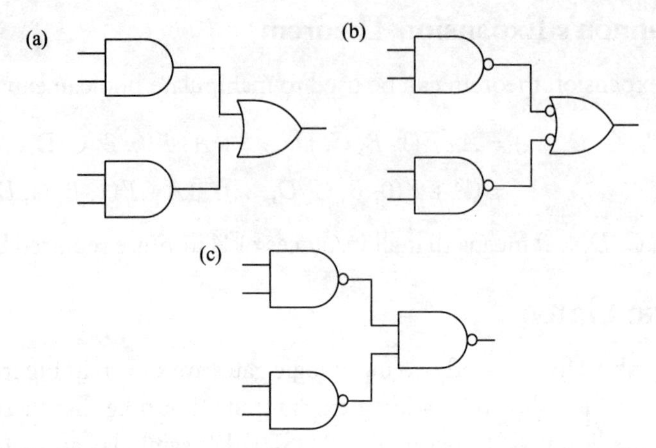

Figure 2.2 Equivalent circuit representations.

of the traditional "spade"-shaped symbols, but are "not preferred" according to IEEE Standard 91-1984. As will be seen in Chapter 3, IEEE notation is useful for describing complex logic blocks, but simple sketches are often clearer if done with the traditional symbols. A circle shows logic inversion. Note that there are two forms of the NAND and NOR gates. From de Morgan's law, it can be seen that the two forms are equivalent in each case.

In drawing circuit diagrams, it is desirable, for clarity, to choose the form of a logic gate that allows inverting circles to be joined. The circuits of Figure 2.2 are identical in function. If the circuit of Figure 2.2(a) is to be implemented using NAND gates, the diagram of Figure 2.2(b) may be preferable to that of Figure 2.2(c) because the function of the circuit is clearer.

2.3 Combinational Logic Design

The values of the output variables of combinational logic are dependent only on the input values and are independent of previous input values or states. Sequential logic, on the other hand, has outputs that depend on the previous states of the system. The design of sequential systems is described in Chapter 6.

The major design objective is usually to minimize the cost of the hardware needed to implement a logic function. That cost can usually be expressed in terms of the number of gates, although for technologies such as programmable logic, there are other limitations, such as the number of terms that may be implemented. Other design objectives may include testability (discussed in detail in Chapter 12) and reliability.

Table 2.7 Minterms and Maxterms

A	B	C	Z	
0	0	0	1	m_0
0	0	1	1	m_1
0	1	0	0	M_2
0	1	1	0	M_3
1	0	0	0	M_4
1	0	1	1	m_5
1	1	0	0	M_6
1	1	1	1	m_7

Before describing the logic design process, some terms have to be defined. In these definitions, it is assumed that we are designing a piece of combinational logic with a number of input variables and a single output.

A *minterm* is a Boolean AND function containing exactly one instance of each input variable or its inverse. A *maxterm* is a Boolean OR function with exactly one instance of each variable or its inverse. For a combinational logic circuit with n input variables, there are 2^n possible minterms and 2^n possible maxterms. If the logic function is true at row i of the standard truth table, that minterm exists and is designated by m_i. If the logic function is false at row i of the standard truth table, that maxterm exists and is designated by M_i. For example, Table 2.7 defines a logic function. The final column shows the minterms and maxterms for the function.

The logic function may be described by the logic OR of its minterms:

$$Z = m_0 + m_1 + m_5 + m_7$$

A function expressed as a logical OR of distinct minterms is in *sum of products* form.

$$Z = \bar{A} \cdot \bar{B} \cdot \bar{C} + \bar{A} \cdot \bar{B} \cdot C + A \cdot \bar{B} \cdot C + A \cdot B \cdot C$$

Each variable is inverted if there is a corresponding 0 in the truth table and not inverted if there is a 1.

Similarly, the logic function may be described by the logical AND of its maxterms.

$$Z = M_2 \cdot M_3 \cdot M_4 \cdot M_6$$

A function expressed as a logical AND of distinct maxterms is in *product of sums* form.

$$Z = (A + \bar{B} + C) \cdot (A + \bar{B} + \bar{C}) \cdot (\bar{A} + B + C) \cdot (\bar{A} + \bar{B} + C)$$

Table 2.8 Truth Table for $Z = A + \bar{A} \cdot \bar{B}$

A	B	Z
0	0	1
0	1	0
1	0	1
1	1	1

Each variable is inverted if there is a corresponding 1 in the truth table and not inverted if there is a 0.

An *implicant* is a term that covers at least one true value and no false values of a function. For example, the function $Z = A + \bar{A} \cdot \bar{B}$ is shown in Table 2.8.

The implicants of this function are $A \cdot B$, A, $\bar{B}$, $\bar{A} \cdot \bar{B}$, $A \cdot \bar{B}$. The non-implicants are $\bar{A}$, B, $\bar{A} \cdot B$.

A *prime implicant* is an implicant that covers one or more minterms of a function, such that the minterms are not all covered by another single implicant. In the example above, A, $\bar{B}$ are prime implicants. The other implicants are all covered by one of the prime implicants. An *essential prime implicant* is a prime implicant that covers an implicant not covered by any other prime implicant. Thus, A, $\bar{B}$ are essential prime implicants.

2.3.1 Logic Minimization

The function of a combinational logic circuit can be described by one or more Boolean expressions. These expressions can be derived from the specification of the system. It is very likely, however, that these expressions are not initially stated in their simplest form. Therefore, if these expressions were directly implemented as logic gates, the amount of hardware required would not be minimal. Therefore, we seek to simplify the Boolean expressions and hence minimize the number of gates needed. Another way of stating this is to say that we are trying to find the set of prime implicants of a function that is necessary to fully describe the function.

In principle, it is possible to simplify Boolean expressions by applying the various rules of Boolean algebra described in Section 2.1. It does not take long, however, to realize that this approach is slow and error prone. Other techniques have to be employed. The technique described here, *Karnaugh maps*, is a graphical method, although it is effectively limited to problems with six or fewer variables. The *Quine-McCluskey* algorithm is a tabular method, which is not limited in the number of variables and is well suited to tackling problems with more than one output. Quine-McCluskey can be performed by hand, but it is generally less easy than the Karnaugh map method. It is better implemented as a computer program. Logic minimization belongs, however, to the *NP-complete* class of problems. This means that as the number of variables increases, the time to find a solution increases exponentially. Therefore, heuristic methods have been developed that find acceptable, but possibly less than optimal, solutions. The *Espresso* program implements heuristic methods that reduce to the Quine-McCluskey algorithm for small problems. Espresso has been used in a number of logic synthesis systems. Therefore, the approach adopted here is to use Karnaugh maps for small problems with a single output and up to six inputs. In general, it makes sense to use an EDA program to solve larger problems.

The Karnaugh map (or K-map, for short) method generates a solution in sum-of-products or product-of-sums form. Such a solution can be implemented directly as two-level AND-OR or OR-AND logic (ignoring the cost of generating the inverse values of inputs). AND-OR logic is equivalent to NAND-NAND logic, and OR-AND logic is equivalent to NOR-NOR logic. Sometimes, a cheaper solution (in terms of the number of gates) can be found by factorizing the two-level, minimized expression to generate more levels of logic—two-level minimization must be performed before any such factorization. Again, we shall assume that if such factorization is to be performed, it will be done using an EDA program, such as *SIS*.

2.3.2 Karnaugh Maps

A Karnaugh map is effectively another way to write a truth table. For example, the Karnaugh map of a general two-input truth table is shown in Figure 2.3.

Similarly, three- and four-input Karnaugh maps are shown in Figures 2.4 and 2.5, respectively. Note that along the top edge of the three-variable Karnaugh map and along both edges of the four-variable map, only one variable changes at a time—the sequence is 00, 01, 11, 10, not the normal binary counting sequence. Hence, for example, the columns in which *A* is true are adjacent. Therefore, the left and right edges, and the top and bottom in the four-variable map, are also adjacent—*B*

A	B	Z
0	0	Z_0
0	1	Z_1
1	0	Z_2
1	1	Z_3

Z:

B \ A	0	1
0	Z_0	Z_2
1	Z_1	Z_3

Figure 2.3 Two-input Karnaugh map.

C \ AB	00	01	11	10
0	Z_0	Z_2	Z_6	Z_4
1	Z_1	Z_3	Z_7	Z_5

Figure 2.4 Three-input Karnaugh map.

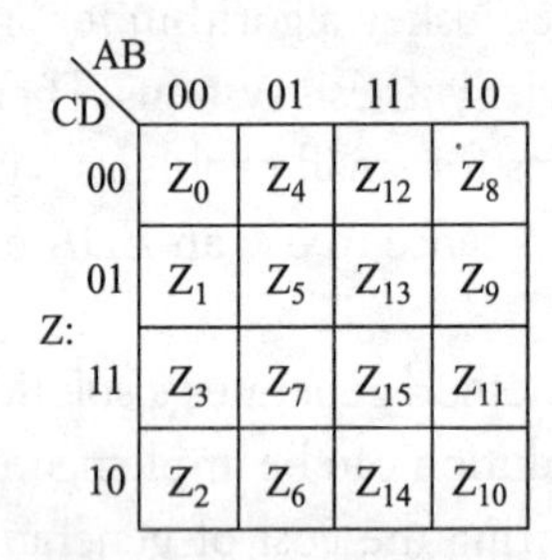

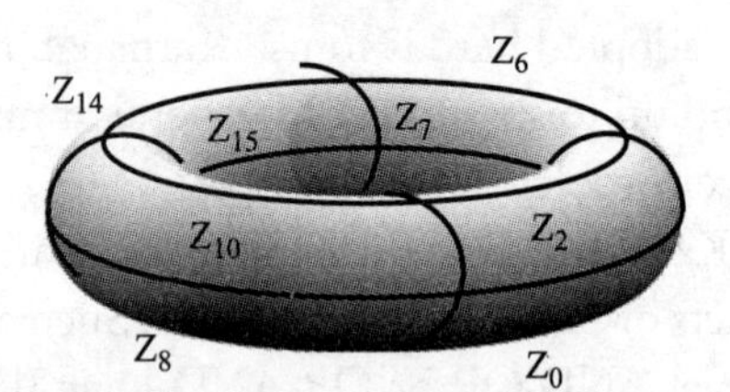

Figure 2.5 Four-input Karnaugh map.

is false in the leftmost and rightmost columns. The three variable map is therefore really a tube, and the four-variable map is a torus, as shown in Figure 2.5. Of course, the maps are drawn as squares for convenience!

A five-variable Karnaugh map is drawn as 2 four-variable maps, one representing the truth table when the fifth variable, E, is false, and the other when E is true. Squares at the same coordinates on both maps are considered to be adjacent. Similarly, a six-variable Karnaugh map is drawn as 4 four-variable maps corresponding to $\bar{E} \cdot \bar{F}$, $\bar{E} \cdot F$, $E \cdot \bar{F}$, and $E \cdot F$, respectively. For this to work, the Karnaugh maps have to be arranged in the pattern as the entries in the two-variable map. Hence, squares at the same location in adjacent maps can be considered adjacent. In practice, therefore, it is not feasible to consider Karnaugh maps with more than six variables.

Implicants can be read from Karnaugh maps by circling groups of $1, 2, 4, 8, \ldots 2^n$ true values. For example, the function $Z = \bar{A} \cdot \bar{B} + \bar{A} \cdot B$ can be expressed as shown in Table 2.9.

Table 2.9 Truth Table for $Z = \bar{A} \cdot \bar{B} + \bar{A} \cdot B$

A	B	Z
0	0	1
0	1	1
1	0	0
1	1	0

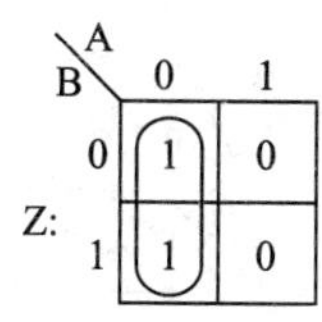

Figure 2.6 Karnaugh map for a two-input function.

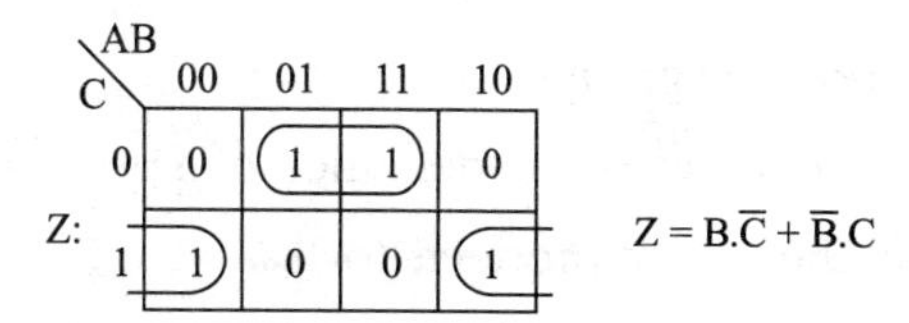

Figure 2.7 Groupings on a three-input Karnaugh map.

The corresponding Karnaugh map is shown in Figure 2.6. We can now circle the two adjacent 1s as shown. This grouping represents the function $Z = \bar{A}$ because it lies in the column $A = 0$, and because within the grouping, B takes both 0 and 1 values and hence we do not care about its value. Therefore, by grouping patterns of 1s, logic functions can be minimized. Examples of three- and four-variable Karnaugh maps are shown in Figures 2.7 and 2.8. In both cases, by considering that the edges of the Karnaugh maps are adjacent, groupings can be made that include 1s at two or four edges.

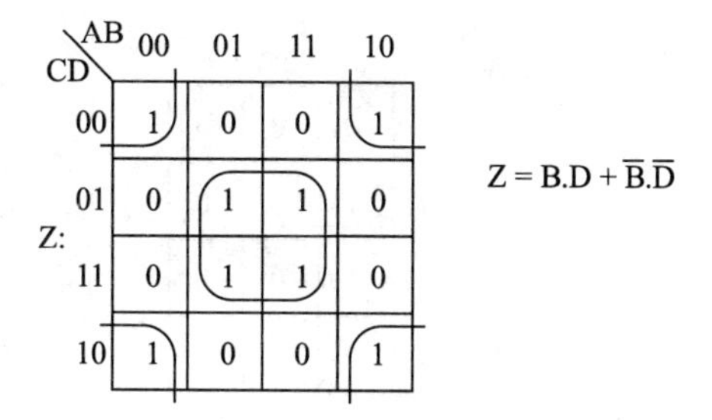

Figure 2.8 Groupings on a four-input Karnaugh map.

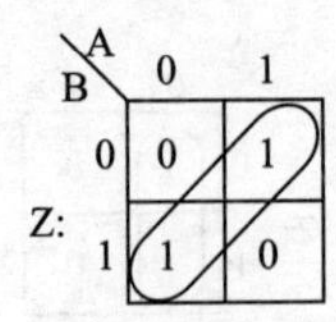

Figure 2.9 Exclusive OR grouping on a Karnaugh map.

The rules for reading *prime implicants* from a Karnaugh map are as follows.

- Circle the largest possible groups.
- Avoid circles inside circles (see the definition of a prime implicant).
- Circle 1s and read the sum of products for Z.
- Circle 0s and read the sum of products for $\bar{Z}$.
- Circle 0s and read the product of sums for Z.
- Circle 1s and read the product of sums for $\bar{Z}$.

Diagonal pairs, as shown in Figure 2.9, correspond to XOR functions.

The Karnaugh map of Figure 2.10 has three prime implicants circled. The function can be read as $Z = B \cdot \bar{C} \cdot D + \bar{A} \cdot C \cdot D + \bar{A} \cdot B \cdot D$. The vertical grouping, shown with a dashed line, covers 1s covered by the other groupings. This grouping is therefore *redundant* and can be omitted. Hence, the function can be read as $Z = B \cdot \bar{C} \cdot D + \bar{A} \cdot C \cdot D$.

Assuming that all the prime implicants have been correctly identified, the minimal form of the function can be read by selecting all the essential prime implicants (i.e., those circles that circle 1s—or 0s—not circled by any other group), together with sufficient other prime implicants needed to cover all the 1s (or 0s). Redundant groupings can be ignored, but under some circumstances it may be desirable to include them.

Z:

CD \ AB	00	01	11	10
00	0	0	0	0
01	0	1	1	0
11	1	1	0	0
10	0	0	0	0

Figure 2.10 Redundant grouping on a Karnaugh map.

A	B	Z
0	0	1
0	1	-
1	0	0
1	1	1

Z:

B \ A	0	1
0	1	0
1	-	1

Figure 2.11 "Don't care" on a Karnaugh map.

Incompletely specified functions have "don't cares" in the truth tables. These don't cares correspond to input combinations that will not (or should not) occur. For example, consider the truth table of Figure 2.11.

The don't care entries can be included or excluded from groups as convenient, in order to get the largest possible groupings, and hence the smallest number of implicants. In the example, we could treat the don't care as a 0 and read $Z = \bar{A} \cdot \bar{B} + A \cdot B$, or treat the don't care as a 1 and read $Z = \bar{A} + B$.

2.4 Timing

The previous section dealt with minimizing Boolean expressions. The minimized Boolean expressions can then be directly implemented as networks of gates or on programmable logic. All gates have a finite delay between a change at an input and a change at an output. If gates are used, therefore, different paths may exist in the network, with different delays. This may cause problems.

To understand the difficulties, it is helpful to draw a *timing diagram*. This is a diagram of the input and output waveforms as a function of time. For example, Figure 2.12 shows the timing diagram for an inverter. Note the stylized (finite) rise

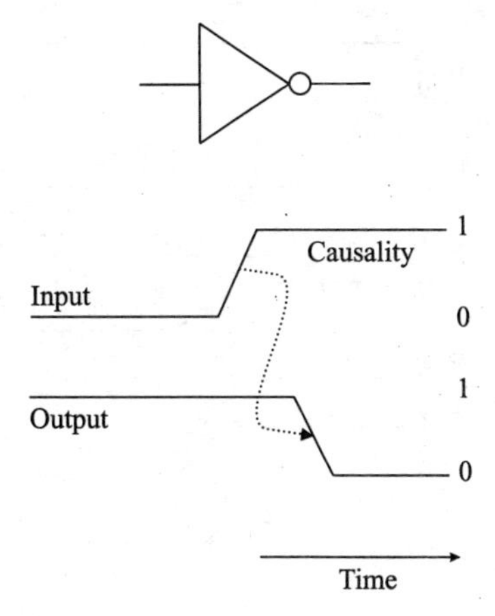

Figure 2.12 Timing diagram for inverter.

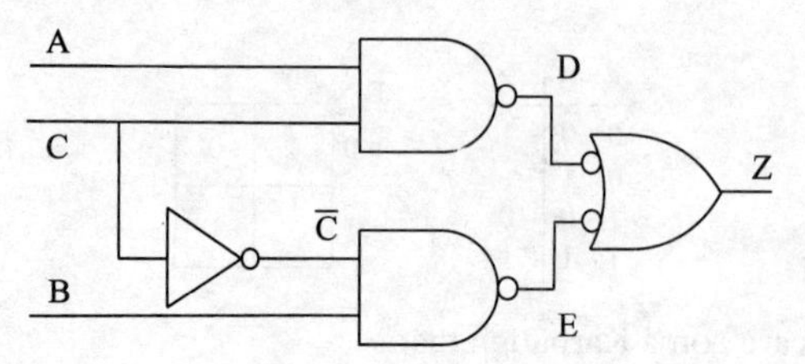

Figure 2.13 Circuit with static 1 hazard.

and fall times. An arrow shows causality, that is, the fact that the change in the output results from a change in the input.

A more complex circuit would implement the function

$$Z = A \cdot C + B \cdot \bar{C}$$

The value of $\bar{C}$ is generated from C by an inverter. A possible implementation of this function is therefore given in Figure 2.13. In practice, the delay through each gate and through each type of gate would be slightly different. For simplicity, however, let us assume that the delay through each gate is one unit of time. To start with, let $A = 1$, $B = 1$. The output, Z, should be at 1 irrespective of the value of C. Let us see, by way of the timing diagram in Figure 2.14, what happens when C changes from 1 to 0. One unit of time after C changes $\bar{C}$ and D change to 1. In turn, these changes cause E and Z to change to 0 another unit of time later. Finally, the change in E causes Z to change back to 1 a further unit of time later. This change

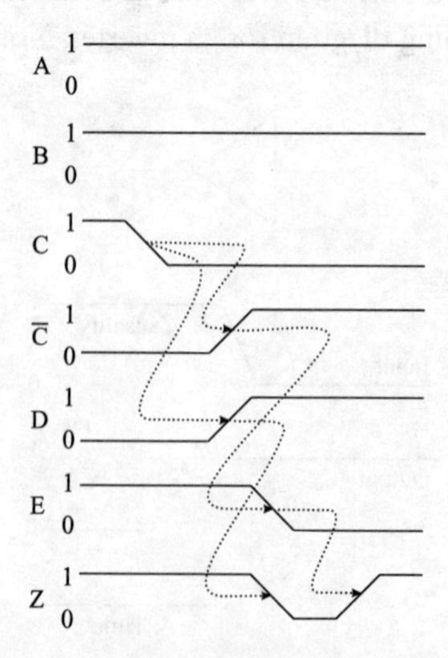

Figure 2.14 Timing diagram for the circuit of Figure 2.13.

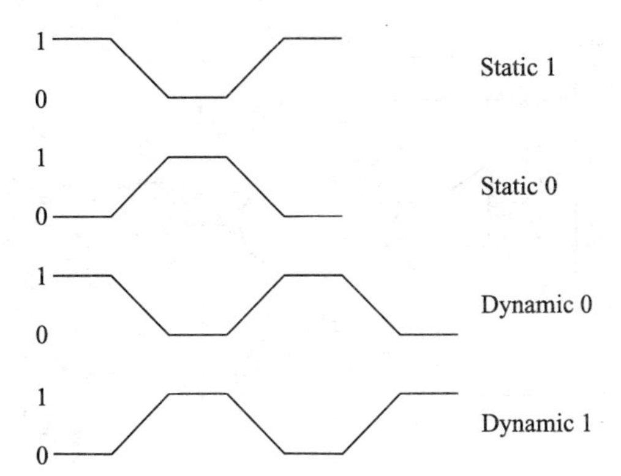

Figure 2.15 Types of hazard.

in Z from 1 to 0 and back to 1 is known as a *hazard.* A hazard occurs as a result of delays in a circuit.

Figure 2.15 shows the different types of hazard that can occur. The hazard in the circuit of Figure 2.13 is a static 1 hazard. Static 1 hazards can only occur in AND-OR or NAND-NAND logic. Static 0 hazards can only occur in OR-AND or NOR-NOR logic. Dynamic hazards do not occur in two-level circuits. They require three or more unequal signal paths. Dynamic hazards are often caused by poor factorization in multi-level minimization.

Static hazards, on the other hand, can be avoided by designing with redundant logic. For example, the Karnaugh map of the circuit function of Figure 2.13 is shown in Figure 2.16. The redundant prime implicant is shown as a dashed circle. The redundant gate corresponding to this prime implicant can be introduced to eliminate the hazard. The circuit function is therefore

$$Z = A \cdot C + B \cdot \bar{C} + A \cdot B$$

The circuit is shown in Figure 2.17. Now, F is independent of C. If $A = B = 1$, $F = 0$. F stays at 0 while C changes; therefore, Z stays at 1. (See Section 11.3.2 for another good reason why circuits with redundancy should be avoided.)

Z:

C \ AB	00	01	11	10
0	0	1	1	0
1	0	0	1	1

Figure 2.16 Redundant term on a Karnaugh map.

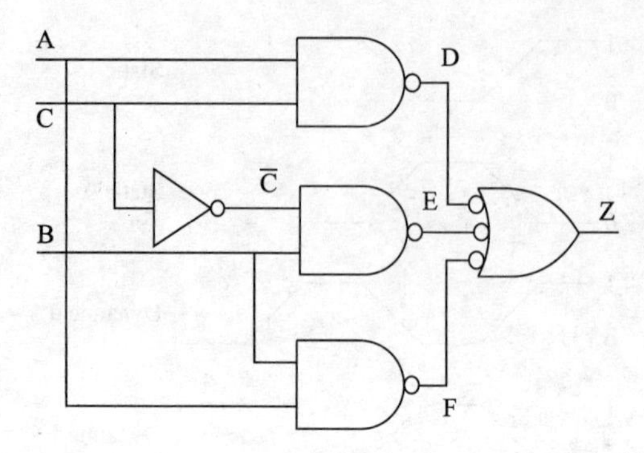

Figure 2.17 Hazard-free circuit.

2.5 Number Codes

Digital signals are either control signals of some kind or information. In general, information takes the form of numbers or characters. These numbers and characters have to be coded in a form suitable for storage and manipulation by digital hardware. Thus, one integer or one character may be represented by a set of bits. From the point of view of a computer or other digital system, no one system of coding is better than another. There do, however, need to be standards, so that different systems can communicate. The standards that have emerged are generally also designed such that a human being can interpret the data if necessary.

2.5.1 Integers

The simplest form of coding is that of positive integers. For example, a set of three bits would allow us to represent the decimal integers 0 to 7. In base 2 arithmetic, 000_2 represents 0_{10}, 011_2 represents 3_{10}, and 111_2 represents 7_{10}. As with decimal notation, the most significant bit is on the left.

For the benefit of human beings, strings of bits may be grouped into sets of three or four and written using *octal* (base 8) or *hexadecimal* (base 16) notation. For example, 66_8 is equal to $110\ 110_2$ or 54_{10}. For hexadecimal notation, the letters A to F represent the decimal numbers 10 to 15. For example, EDA_{16} is $1110\ 1101\ 1010_2$ or 7332_8 or 3802_{10}.

The simple translation of a decimal number into bits is sufficient for zero and positive integers. Negative integers require additional information. The simplest approach is to set aside one bit as a sign bit. Therefore, $0\ 110_2$ might represent $+6_{10}$, while $1\ 110_2$ would represent -6_{10}. While this makes translation between binary and decimal numbers simple, the arithmetic operations of addition and subtraction require that the sign bits be checked before an operation can be performed on two numbers. It is common, therefore, to use a different notation for signed integers: *two's complement*. The principle of two's complement notation is that the code for $-b$, where b is a binary number represented using n bits, is the code given by $2^n - b$. For example, -6_{10} is represented by $10000_2 - 0110_2$, which is 1010_2. The same result is obtained by inverting all the bits and adding 1: -6_{10} is $1001_2 + 1 = 1010_2$.

The advantage of two's complement notation is that addition and subtraction may be performed using exactly the same hardware as for unsigned arithmetic; no sign checking is needed. The major disadvantage is that multiplication and division become much more complicated. Booth's algorithm, described in Section 5.7, is a technique for multiplying two's complement numbers.

2.5.2 Fixed Point Numbers

For many applications, non-integer data needs to be stored and manipulated. The binary representation of a *fixed-point* number is exactly the same as for an integer number, except that there is an implicit "decimal" point. For example, 6.25 is equal to $2^2 + 2^1 + 2^{-2}$ or 110.01_2. Instead of representing the point, the number 11001_2 (25_{10}) is stored with the implicit knowledge that it and the results of any operations involving it have to be divided by 2^2 to obtain the true value. Notice that all operations, including two's complement representations, are the same as for integer numbers.

2.5.3 Floating-Point Numbers

The number of bits that have been allocated to represent fractions limits the range of fixed point numbers. *Floating-point* numbers allow a much wider range of accuracy. In general, floating-point operations are only performed using specialized hardware because they are very computationally expensive. A typical *single precision* floating-point number has 32 bits, of which 1 is the sign bit (s), 8 are the exponent (e), biased by an offset ($2^e - 1 = 127$), and the remaining 23 are the mantissa (m), such that a decimal number is represented as

$$(-1)^s \times 1 \cdot m \times 2^e$$

IEEE standard 754-1985 defines formats for 32-, 64-, and 128-bit floating-point numbers, with special patterns for $\pm\infty$ and the results of invalid operations, such as $\sqrt{-1}$.

2.5.4 Alphanumeric Characters

Characters are commonly represented by 7 or 8 bits. The ASCII code is widely used. Seven bits allow the basic Latin alphabet in upper and lower cases, together with various punctuation symbols and control codes to be represented. For example, the letter A is represented by 1000001. For accented characters, 8-bit codes are commonly used. Manipulation of text is normally performed using general purpose computers rather than specialized digital hardware. Non-European languages may use 16 or 32 bits to represent individual characters.

2.5.5 Gray Codes

In the normal binary counting sequence, the transition from 0111 (7_{10}) to 1000 (8_{10}) causes 3 bits to change. In some circumstances, it may be undesirable that several bits should change at once because the bit changes may not occur at exactly the same time. The intermediate values might generate spurious warnings. A *Gray* code is one in which only 1 bit changes at a time. For example a 3-bit Gray code would count through the following sequence (other Gray codes can also be derived):

000
001
011
010
110
111
101
100

Note that the sequence of bits on a K-map is a Gray code. Another application of Gray codes is as a position encoder on a rotating shaft, as shown in Figure 2.18. Only 1-bit changes at each transition, so missed transitions are easily spotted.

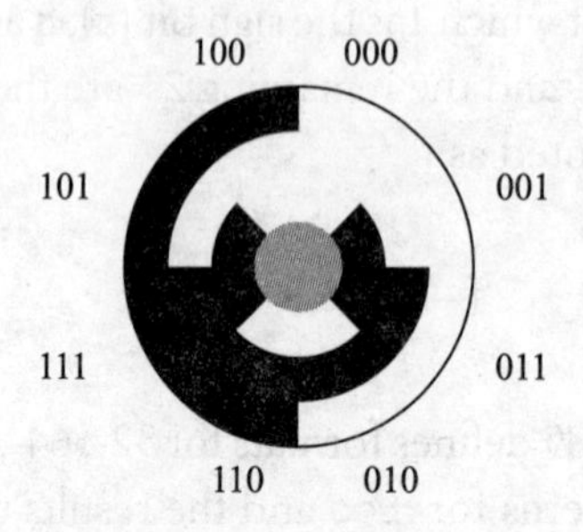

Figure 2.18 Gray code as shaft encoder.

Table 2.10 Two-Dimensional Parity

	Bit 7	Bit 6	Bit 5	Bit 4	Bit 3	Bit 2	Bit 1	Bit 0	Parity
Word 0	0	1	0	1	0	1	1	0	0
Word 1	0	1	0	0	1	0	0	0	0
Word 2	0	1	0	1	0	0	1	1	0
Word 3	0	1	0	0	1	0	0	1	1
Word 4	0	1	0	0	0	0	1	1	1
Word 5	0	1	0	0	1	0	0	0	0
Word 6	0	1	0	0	0	1	0	0	0
Word 7	0	1	0	0	1	1	0	0	1
Parity	0	0	0	0	0	1	1	1	1

2.5.6 Parity Bits

When data are transmitted either by wire or by using radio communications, there is always the possibility that noise may cause a bit to be misinterpreted. At the very least, it is desirable to know that an error has occurred, and it may be desirable to transmit sufficient information to allow any error to be corrected.

The simplest form of error detection is to use a parity bit with each word of data. For each 8 bits of data, a ninth bit is sent that is 0 if there is an even number of 1s in the data word (even parity) or 1 otherwise. Alternatively, odd parity can be used; in which case, the parity bit is inverted. This is sufficient if the chances of an error occurring are low. We cannot tell which bit is in error, but knowing that an error has occurred means that the data can be transmitted again. Unfortunately, if two errors occur, the parity bit might appear to be correct. A single error can be corrected by using a two-dimensional parity scheme, in which every ninth word is itself a set of parity bits, as shown in Table 2.10. If a single error occurs, both the row parity and column parity will be incorrect, allowing the erroneous bit to be identified and corrected. Certain multiple errors are also detectable and correctable.

By using a greater number of parity bits, each derived from part of the word, multiple errors can be detected and corrected. The simplest forms of such codes were derived by Hamming in 1948. Better codes were derived by Reed and Solomon in 1960.

Summary

Digital design is based on Boolean algebra. The rules of Boolean algebra allow logical expressions to be simplified. The basic logical operators can be implemented as digital building blocks—gates. Graphical methods, such as Karnaugh maps, are suitable tools for finding the minimal forms of Boolean expressions with fewer than six variables. Larger problems can be tackled with computer-based methods. Gates have delays, which means that non-minimal forms of Boolean expressions may be needed to prevent timing problems, known as hazards. Data can be represented using sets of bits. Different types of data can be encoded to allow manipulation. Error detecting codes are used when data is transmitted over radio or other networks.

Further Reading

The principles of Boolean algebra and Boolean minimization are covered in many books on digital design. Recommended are those by Wakerly [25] and Hill and Peterson [6]. De Micheli [10] describes the Espresso algorithm, which sits at the heart of many logic optimization software packages. Espresso may be downloaded from www-cad.eecs.berkeley.edu/.

Error detection and correction codes are widely used in communications systems. Descriptions of these codes can be found in, for example, Hamming [8].

Exercises

2.1 Derive Boolean expressions for the circuits of Figure 2.19; use truth tables to discover if they are equivalent.

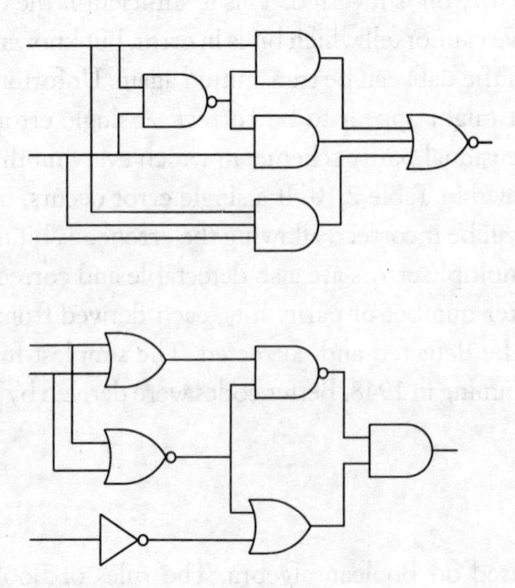

Figure 2.19 Circuits for Exercise 2.1.

2.2 Minimize

(a) $Z = m_0 + m_1 + m_2 + m_5 + m_7 + m_8 + m_{10} + m_{14} + m_{15}$

(b) $Z = m_3 + m_4 + m_5 + m_7 + m_9 + m_{13} + m_{14} + m_{15}$

2.3 Describe two ways of representing negative binary numbers. What are the advantages and disadvantages of each method?

2.4 A floating-point decimal number may be represented as:

$$(-1)^s \times 1 \cdot m \times 2^e$$

Explain what the binary numbers s, m, and e represent. How many bits would typically be used for s, m, and e in a single precision floating-point number?

3 Combinational Logic Using SystemVerilog Gate Models

Combinational logic is stateless: Changes in inputs are immediately reflected by changes in outputs. In this chapter, we introduce the basic ideas of modeling in SystemVerilog by looking at combinational logic described in terms of gates.

3.1 Modules and Files

The basic unit of a SystemVerilog design is the **`module`**. For example, a two-input AND gate might be described by:

```
module And2 (output wire z, input wire x, y);

assign z = x & y;

endmodule
```

The words shown in **bold** are keywords. The module description starts with the keyword **`module`**, followed by the name of the module and a list of inputs and outputs in parentheses. The module finishes with the keyword **`endmodule`**. Note that a semicolon (";") follows the module header, but that there is no semicolon following the **`endmodule`** keyword.

The header contains the inputs and outputs of the module. Here, one **`output`** of type **`wire`** is declared, `z`, followed by `x` and `y`, defined by the keywords **`input`**

Table 3.1 Arithmetic and Logical Operators in Order of Decreasing Precedence

<table>
<tr><th>Arithmetic</th><th>Bitwise</th><th>Logical</th></tr>
<tr><td>+ – (unary)</td><td>~</td><td>!</td></tr>
<tr><td>* / %</td><td></td><td></td></tr>
<tr><td>+ –</td><td></td><td></td></tr>
<tr><td></td><td><< >></td><td></td></tr>
<tr><td></td><td>&</td><td></td></tr>
<tr><td></td><td>^ ^~</td><td></td></tr>
<tr><td></td><td>|</td><td></td></tr>
<tr><td></td><td></td><td>&&</td></tr>
<tr><td></td><td></td><td>||</td></tr>
</table>

and **`wire`**. Inputs and outputs can appear in any order; the convention with SystemVerilog gates is that outputs are declared before inputs. Because `x` and `y` have the same direction and the same type, they can be listed together after the keywords.

In this example, the model has only one statement. The keyword **`assign`** is used to indicate a so-called *continuous assignment*—this will be explained later. The bitwise AND of `x` and `y` is assigned to `z`. Arithmetic and logical operators in SystemVerilog are based on those in C. A full list is given in Table 3.1.

The entire module can be contained in single file. It is possible to have more than one module in a file, but this is not advisable because any change to one module requires compilation of the entire file.

It is recommended that you follow these guidelines when organizing your work.

- Put each module in a separate file.
- The file name and the module name should be the same; give the file name the extension ".v" (for Verilog) or ".sv" (for SystemVerilog). (For the examples in this book, the extension ".v" is used.)
- Do not use spaces in file names or folders/directories. (Some tools have difficulties, even when this is allowed by the operating system.)

3.2 Identifiers, Spaces, and Comments

SystemVerilog is case sensitive (like C). Keywords must be lower case. Identifiers (such as "And2") may be mixed case. It is strongly recommended that usual software engineering rules about identifiers should be applied.

- Meaningful, non-cryptic names should be used, based on English words.
- Use mixed-case with consistent use of case.
- Don't use excessively long identifiers (15 characters or fewer).
- Don't use identifiers that may be confused (e.g., two identifiers that differ by an underscore).
- Identifiers may consist of letters, numbers, and underscores ("_"), but the first character must not be a number.
- System tasks and functions start with a dollar symbol ("$").
- It is possible to include other symbols in identifiers by starting the identifier with a backslash ("\"). This is intended for transferring data between tools, so use with extreme caution!

White space (spaces, carriage returns) should be used to make models more readable. There is no difference between one white space character and many.

Comments may be included in a SystemVerilog description by putting two slashes on a line ("//"). All text between the slashes and the end of the line is ignored. This is similar to the C++ style of comment. There is also a C-style block comment ("/*...*/") in SystemVerilog. It is strongly recommended that comments should be included to aid in the understanding of SystemVerilog code. Each SystemVerilog file should include a header, which typically contains

- The name(s) of the design units in the file
- File name
- A description of the code
- Limitations and known errors
- Any operating system and tool dependencies
- The author(s), including a full address
- A revision number

For example:

```
//////////////////////////////////////////////////////////
// Design unit   : And2
//               :
// File name     : And2.v
//               :
// Description   : Model of basic 2 input AND
//               : gate. Inputs of type wire.
```

```
//                  :
// Limitations      : None
//                  :
// System           : IEEE 1800-2005
//                  :
// Author           : Mark Zwolinski
//                  : School of Electronics and Computer
//                  : Science
//                  : University of Southampton
//                  : Southampton SO17 1BJ, UK
//                  : mz@ecs.soton.ac.uk
//
// Revision         : Version 1.0 04/02/09
/////////////////////////////////////////////////////
```

3.3 Basic Gate Models

Built into SystemVerilog are a number of low-level gate primitives. These include:

and, **or**, **nand**, **nor**, **xor**, **xnor**, **not**, **buf**.

These are keywords, which will be shown in bold font. SystemVerilog is case sensitive; keywords are always lower case. The meaning of the gates is probably self-evident: **xnor** is an exclusive NOR gate (in other words, the output is true if the inputs are equal); **buf** is a non-inverting buffer. There are several other primitives, but these are sufficient for our purposes.

To distinguish one instance of a gate from another, a label follows the gate primitive name (see the following for an example).

A gate is connected to one or more nets. These nets are listed in parentheses. The convention is that the output comes first and is followed by the input(s). A NAND gate with inputs a and b and output y might appear in a piece of SystemVerilog as:

```
nand g1 (y, a, b);
```

where g1 is the label for that gate. Note the semicolon (;) at the end of the instance. White space is not important, so this description could be split over two or more lines, or formatted to line up with other statements.

It is possible to have more than one gate instance declared at the same time:

```
nand g1 (y, a, b), g2 (w, c, d);
```

This describes two gates (g1 and g2). This can be split over two or more lines. While legal, this is not really recommended because it can make circuit descriptions difficult to read.

There are two further pieces of information that can be declared with a gate: the signal strength and the delay. In CMOS circuits, signal strength is not usually modeled, other than in the case of a high impedance state. We will discuss delay modeling after we have looked at the structure of a netlist description.

3.4 A Simple Netlist

A netlist is a list of nets (!) and the gates (or other elements) that connect them. Let us see how a simple combinational logic circuit can be described.

Figure 3.1 shows a simple combinational logic circuit, with one output (y), three inputs (a, b, c), and three internal nodes (d, e, f). The gates and inverter are labeled (g1, g2, g3, g4). This is a SystemVerilog description of the circuit:

```
module ex1 (output wire y, input wire a, b, c);
  wire d, e, f;
  not g1 (d, c);
  and g2 (e, a, c);
  and g3 (f, d, b);
  or g4 (y, e, f);
endmodule
```

The description begins with the keyword **module**, followed by a name. The inputs and outputs are then listed. We will follow the convention used in gate primitives and list the output(s) before the input(s). It is also possible to have bidirectional connections, declared with the **inout** keyword. All the inputs and outputs are declared to be *nets* with the keyword **wire**. In fact, this keyword is not needed, but it is *strongly recommended*, however, that you declare all nets using the **wire** keyword (or the **logic** keyword, as we will see in later chapters).

The second line declares the internal nets of the circuit. Again, the declaration is not strictly needed because once a net is used in a gate description, it is automatically declared. Again, it is recommended that you declare all nets for clarity.

The next four lines are the gate declarations, which we have already discussed. Finally, the end of the description is marked by the keyword **endmodule**.

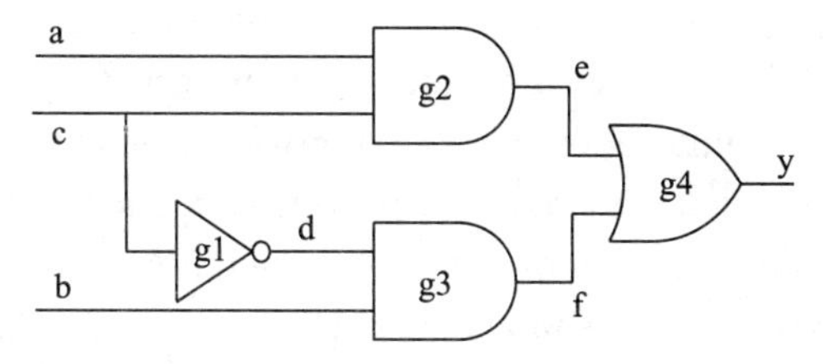

Figure 3.1 Simple combinational circuit.

3.5 Logic Values

In the preceding description, we mentioned logic values and referred briefly to a high impedance state. SystemVerilog allows wires to take four possible values: 0, 1, x (unknown), and z (high impedance). In general, logic gates are designed to generate 0 or 1 at the outputs. x usually indicates some kind of anomalous situation—perhaps an uninitialized flip-flop or a wire that is being driven to two different values by two gates simultaneously.

The high-impedance state, z, is used to model the output of three-state buffers. The purpose of three-state buffers is to allow the outputs of gates to be connected together to form buses, for example. The x state is normally generated when different outputs from two gates are connected together. We would expect, however, that a 1 and a z (or a 0 and a z) driving the same wire would resolve to a 1 (or a 0). Clearly, therefore, not all logic values are equal.

The unknown and high-impedance states can be written as lower case ("x" and "z") or upper case ("X" and "Z") characters. The question mark ("?") can be used as an alternative to the high-impedance state.

3.6 Continuous Assignments

The two-input AND gate at the beginning of the chapter was written using a *continuous assignment*. In general, continuous assignments are used to **`assign`** values to nets. In later chapters, we will see that **`always_comb`** and **`always_ff`** procedural blocks are more useful for describing synthesizable hardware. Continuous assignments are, on the other hand, the most convenient way to describe three-state buffers and to model delays in combinational logic. Three-state buffers will be discussed in more detail in the next chapter. This is an appropriate point, however, to discuss SystemVerilog operators.

3.6.1 SystemVerilog Operators

Most of the arithmetic and logical operators in SystemVerilog are the same as those in C. The arithmetic and logical operators are given in decreasing order of precedence in Table 3.1. The standard arithmetic operators should not need further explanation. % is a modulus operator; ~ is a bitwise negation; ! is a logic negation; << and >> mean shift left and right, respectively; & means AND; | means OR; and ^ means XOR. The meaning of combined operators should be apparent. The bitwise operators can be used as unary reduction operators, taking all the bits in a vector as inputs and giving a single bit output.

For single bits, there is no difference between the bitwise and logical operators. For integers or vectors of more than one bit, the bitwise operators are applied to each bit, while the logical operators apply to the vector as a whole (a non-zero value is true; zero is false). The bitwise operators can be used to construct Boolean functions. For example, an AND-OR-Invert function could be written as:

```
assign y = ~(a & b | c & d);
```

The conditional operator of C is also implemented. This is particularly useful for three-state buffers:

```
assign y = enable ? a : 'z;
```

The notation `'z` means one or more z values. The exact number depends on the target on the left-hand side of the assignment. Here only one z value is assigned. This is a useful shorthand, introduced in SystemVerilog. Alternatively, the exact number of bits can be specified, for example `1'bz`, where `1'b` means "1 bit".

SystemVerilog also has three types of equality operators: ==, ===, and ==?, together with the inverses, !=, !==, and !=?. The differences between these operators are in the ways in which they treat x and z values. The C-like == and != return an unknown value (`1'bx`) if any bit in either operand is x or z. === and !== do an exact comparison between bit patterns, matching x and z values. ==? and !=? treat x and z values in the *right-hand* operand as don't cares. In practice, == and != are synthesizable and === and !== are not synthesizable (because it is not possible to detect x and z values in real hardware). ==? and !=? are synthesizable if the right-hand operand is a constant; for example,

```
assign even = (a ==? 4'b???0);
```

(As noted previously, "?" is an alternative to "z".)

3.7 Delays

While it is possible to design circuits at the gate level, it does make the use of an HDL like SystemVerilog a little pointless. Indeed, it could be argued that writing a netlist by hand is a waste of time. If you are working at that level, you will probably have had to sketch the circuit diagram. So why not use a schematic capture program to draw the circuit, and then generate the netlist from the schematic automatically?

This does not mean you will never encounter a netlist. Another way of generating a netlist is from a synthesis tool or by extracting the circuit *after* physical layout of the circuit. In both of these cases, you will probably wish to verify your design by simulation. You can simply verify the logical functionality of the circuit, but it is

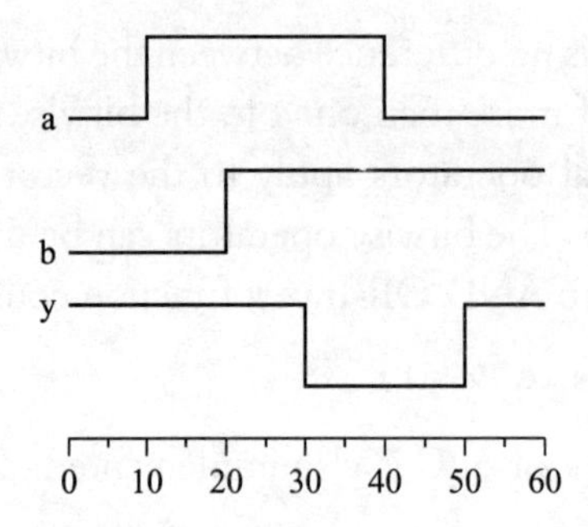

Figure 3.2 NAND function with delay.

often more important to verify that the circuit will work correctly at the normal operating speed. To verify the timing, the simulation model must include timing information. The simplest way to include this information is to model the delay through each gate.[1] For example, a delay of 10 ps through a NAND gate would be written as:

```
nand #10ps g1 (y, a, b);
```

The hash symbol (#) is used to denote a parameter. We will see further examples of parameters in later chapters. Notice that the delay parameter is placed between the type of gate (**nand**) and the name of the instance (g1).

In the previous example, there is one delay parameter. In the case of a NAND gate, the output is at logic 1 if either or both inputs are at logic 0. Therefore, the output will only go to logic 0 after the second of the two inputs has gone to 1. This change will be delayed by 10 ps.

In Figure 3.2, signal b goes to 1 at time 20 ps; signal a goes back to 0 at time 40 ps. Therefore, the pulse on y is 20 ps wide, delayed by 10 ps.

Suppose that a changes back to 0 at time 35 ps. This would suggest that a pulse 5 ps wide would appear at y, again delayed by 10 ps. In fact, the delay has a second meaning: Any pulse less than 10 ps wide is suppressed, Figure 3.3.

This is known as an *inertial delay*. Hence, a pulse is suppressed by *inertial cancellation*.

---

1. In practice, a separate SDF (standard delay format) file, containing the timing information for each gate, would be generated. See Section 10.5.1, but the principle still applies.

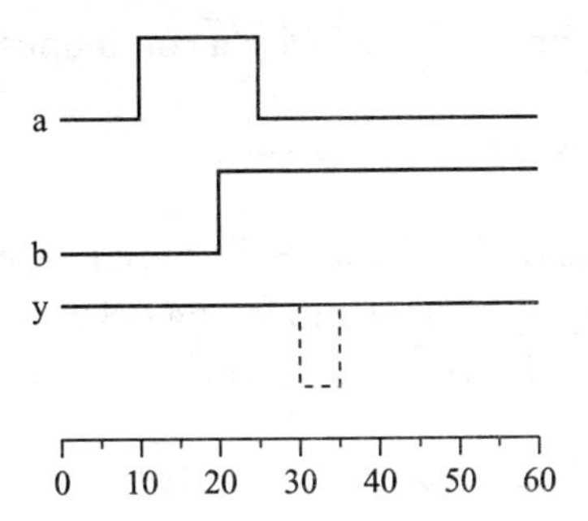

Figure 3.3 NAND function with inertial cancellation.

This delay model assumes that the delay through a gate is the same for a 0 to 1 transition on the output as for a 1 to 0 transition. This assumption is probably valid for CMOS technologies, but may not be true in other cases. If the 0 to 1 and 1 to 0 delays differ, the two values may be specified. For example,

```
nand #(10ps, 12ps) g1 (y, a, b);
```

describes a NAND gate that has a 10 ps delay when the output changes to 1 (rise delay) and a 12 ps delay when the output changes to 0 (fall delay). It is also possible to specify a third delay for the case when the output changes to a high-impedance state.

We can take delay modeling one step further to describe uncertainty. When a circuit is extracted from a silicon layout, it is not possible to exactly predict the delay through each gate or between each gate because of process variations. It is reasonable, however, to expect that the minimum, typical, and maximum delays through a gate can be the minimum, typical, and maximum delays respectively; for example,

```
nand #(8ps:10ps:12ps, 10ps:12ps:14ps) g1 (y, a, b);
```

describes a NAND gate that has a minimum rise delay of 8 ps, a typical rise delay of 10 ps, and a maximum rise delay of 12 ps. Similarly, the fall delay has three values. A simulation can therefore be performed using the minimum, typical, or maximum delays for all gates. In principle, the functionality of the circuit can therefore be verified under extremes of fabrication.

Similarly, a delay can be associated with a continuous assignment:

```
assign #10ps z = x & y;
```

This is an inertial delay that behaves in exactly the same way as for a gate primitive. By the same analogy, rising and falling delays and minimum, typical, and maximum delays can be included, as for a gate:

```
assign #(8ps:10ps:12ps, 10ps:12ps:14ps) z = x & y;
```

3.8 Parameters

The statement:

```
assign #5ps z = x & y;
```

defines the exact delay for an AND gate. Different technologies, and indeed different instances, will have different delays. We could declare a number of alternative modules for an AND gate, each with a different delay. It would be better to write the statement as:

```
assign #delay z = x & y;
```

and to define delay as a **parameter** to the SystemVerilog model.

```
module And2 #(parameter delay) (output wire z,
                                input wire x, y);

  assign #delay z = x & y;
endmodule
```

When the gate is used in a netlist, a value is passed to the model as a parameter:

```
And2 #(5ps) g2  (p, b, q);
```

3.9 Testbenches

If we wish to simulate our circuit to verify that it really does work as expected, we need to apply some test stimuli. We could, of course, write out some test vectors and apply them one at a time, or, more conveniently, write the test data in SystemVerilogThis type of SystemVerilog model is often known as a testbench. Testbenches have a distinctive style. A testbench for a two-input AND gate is shown.

```
module TestAnd2;

  wire a,b,c;

  And2 g1 (c, a, b);

initial
  begin
  a = '0;
  b = '0;
  #100ps a = '1;
  #50ps b = '1;
  end

endmodule
```

Because this is a testbench, that is, a description of the entire world that affects the model we are testing, there are no inputs or outputs to the module. This is characteristic of testbenches. Nets corresponding to the input and output ports of the circuit are first declared. The description consists of an instance of the circuit we are testing, together with a set of input stimuli.

In the instantiation of the `And2` gate, the nets in the testbench are associated with the nets in the gate model, according to the order in which they are written. Thus, `c` is connected to net `z` of the gate, and so on.

An **`initial`**[2] procedural block is declared, in which the initial values of `a` and `b` are defined.[3] After a delay of 100 ps, `a` is updated and 50 ps later `b` is updated. Note the use of **`begin`** and **`end`** to bracket multiple statements.

This is a very simple example of a testbench. It provides sufficient inputs to run a simulation, but the designer would need to look at the simulation results to check that the circuit was functioning as intended. SystemVerilog has the richness of a programming language. Therefore, a testbench could be written to check simulation results against a file of expected responses or to compare two versions of the same circuit.

---

2. Beware. An **`initial`** block is executed once. A net should only be assigned a value from one block. Therefore, it is a mistake to use an **`initial`** block to *initialize* a net and to assign another value elsewhere.

3. This works in SystemVerilogÚsing an initial block to assign a value to a **`wire`** will not work in earlier versions of Verilog.

Summary

In this chapter, we discussed the modeling of logic circuits as netlists of primitive components.

A circuit block begins with the keyword **module** and ends with the keyword **endmodule**. Inputs and outputs are listed in parentheses after the **module** keyword as **input**s, **output**s or **inout**s. Internal **wire**s may optionally be declared. It is good practice to do this as it increases the readability of a design. A number of gate primitives exist in SystemVerilogĠenerally, the connections to these gates are in the order: output(s), input(s), control. Primitives may have one or two (or in the case of three-state primitives, three) delay parameters. The delays are listed following a hash (#). Delays are inertial. Delays may be specified with minimum, typical, and maximum values.

SystemVerilog signals can take four values: 1, 0, x, or z. z has a lower strength than the other values.

Further Reading

The definition of SystemVerilog is contained in the standard IEEE 1800-2005 [2]. This can be bought from the IEEE. There are a number of Verilog books available.

Exercises

3.1 Write a description of a three-input NAND gate with a delay of 5 ps using a continuous assignment.

3.2 Write a description of a three-input NAND gate with a parameterizable delay using a continuous assignment.

3.3 A full adder has the truth table of Table 3.2 for its sum (S) and carry (Co) outputs, in terms of its inputs, A, B and carry in (Ci).
Derive expressions for S and Co using only AND and OR operators. Hence, write a SystemVerilog description of a full adder as a netlist of AND and OR gates and inverters. Do not include any gate delays in your models.

3.4 Write a SystemVerilog testbench to test all combinations of inputs to the full adder of Exercise 3.3. Verify the correctness of your full adder and of the testbench using a SystemVerilog simulator.

Table 3.2 Truth Table for Exercise 3.3

A	B	Ci	S	Co
0	0	0	0	0
0	0	1	1	0
0	1	0	1	0
0	1	1	0	1
1	0	0	1	0
1	0	1	0	1
1	1	0	0	1
1	1	1	1	1

3.5 Modify the gate models of Exercise 3.3 such that each gate has a delay of 1 ns. What is the maximum delay through your full adder? Verify this delay by simulation.

4 Combinational Building Blocks

While it is possible to design all combinational (and indeed sequential) circuits in terms of logic gates, in practice this would be extremely tedious. It is far more efficient, in terms of both the designer's time and the use of programmable logic resources, to use higher level building blocks. If we were to build systems using TTL or CMOS integrated circuits on a printed circuit board, we would look in a catalog and choose devices to implement standard circuit functions. If we use SystemVerilog and programmable logic, we are not constrained to using just those devices in the catalog, but we still think in terms of the same kinds of circuit functions. In this chapter, we look at a number of combinational circuit functions. As we do so, various features of SystemVerilog will be introduced. In addition, the IEEE dependency notation will also be introduced, allowing us to describe circuits using both graphical and textual representations.

4.1 Multiplexers

4.1.1 2 to 1 Multiplexer

A multiplexer can be used to switch one of many inputs to a single output. Typically, multiplexers are used to allow large, complex pieces of hardware to be reused. The IEEE symbol for a 2 to 1 multiplexer is given in Figure 4.1. G is a select symbol.

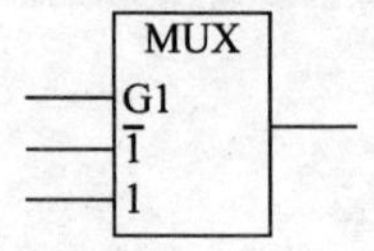

Figure 4.1 2 to 1 multiplexer.

If G is true, the input labeled 1 is connected to the output; if G is false, the input labeled $\bar{1}$ is chosen.

A SystemVerilog model of this multiplexer follows.

```
module mux2 (output logic y,
             input logic a, b, s);

always_comb
  if (s)
    y = b;
  else
    y = a;

endmodule
```

always_comb is a SystemVerilog variant on the general purpose Verilog **always** construct. We will see other variants in later chapters. A procedural **always** block allows various procedural programming constructs to be used. **always_comb** indicates that the block models purely combinational logic at the register transfer level. In a general purpose Verilog **always** block, every input used by that block must be listed. For an **always_comb** block, the inputs are derived automatically. Before SystemVerilog, two of the most common errors made in writing RTL Verilog were the accidental creation of sequential logic and the accidental omission of input signals, resulting in a mismatch between simulated and synthesized behavior. Therefore, it is very strongly recommended that synthesizable combinational logic should always be modeled using the **always_comb** construct.

The **always_comb** block here contains exactly one statement. If more than one statement is needed, they should be grouped using **begin** and **end**. The **if** statement here is self-explanatory: if the select input, s, is true, the logic value at input b is assigned to output y; if false, the value at a is assigned. An **if** statement must occur within a procedural block. Note that the assignment is indicated using a single equals sign (=). This is known as a *blocking* assignment. The significance of this will be explained later. It is sufficient to note here that combinational logic should always be modeled using blocking assignments.

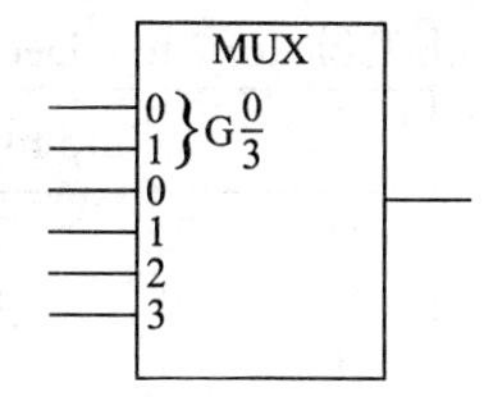

Figure 4.2 4 to 1 multiplexer.

4.1.2 4 to 1 Multiplexer

The symbol for a 4 to 1 multiplexer is shown in Figure 4.2. As before, G is a select symbol. $\frac{0}{3}$ is not a fraction, but means 0-3. Therefore, the binary value on the top two inputs is used to select one of the inputs 0-3.

The 2 to 1 multiplexer model can be extended to a 4 to 1 multiplexer by nesting if statements.

```
module mux4 (output logic y,
             input logic a, b, c, d, s0, s1);

always_comb
  if (s0)
    if (s1)
       y = d;
    else
       y = c;
  else
    if (s1)
       y = b;
    else
       y = a;

endmodule
```

4.2 Decoders

4.2.1 2 to 4 Decoder

A decoder converts data that has previously been encoded into some other form. For example, n bits can represent 2^n distinct values. The truth table for a 2 to 4 decoder is given in Table 4.1.

Table 4.1 Truth Table for 2 to 4 Decoder

Inputs		Outputs			
A1	A0	Z3	Z2	Z1	Z0
0	0	0	0	0	1
0	1	0	0	1	0
1	0	0	1	0	0
1	1	1	0	0	0

The IEEE symbol for a 2 to 4 decoder is shown in Figure 4.3. BIN/1-OF-4 indicates a binary decoder in which one of four outputs will be asserted. The numbers give the "weight" of each input or output.

We could choose to treat each of the inputs and outputs separately, but as they are obviously related, it makes sense to treat the input and output as two vectors of size 2 and 4, respectively. Vectors can be described using an array of variables; for example:

```
logic [3:0] four_bit_array;
```

The 2 to 4 decoder can be modeled using a **case** statement:

```
module decoder (output logic [3:0] y,
                input logic [1:0] a);

always_comb
  case (a)
    0 : y = 1;
    1 : y = 2;
    2 : y = 4;
    3 : y = 8;
    default : y = 'x;
  endcase
endmodule
```

Depending on the numerical value of a, one of the branches in the case statement is selected. There is, however, a sleight of hand here. SystemVerilog is not a strongly typed language. The input, a, is declared to be a 2-bit variable of type logic, but it is interpreted as an integer. This is acceptable in SystemVerilog, but would be

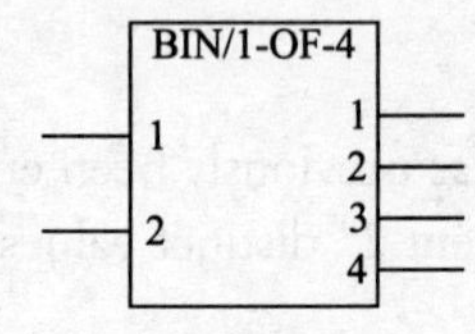

Figure 4.3 2 to 4 decoder.

completely illegal in many other HDLs and programming languages. This ability to interpret bit patterns automatically is very powerful, but can be dangerous. The bit pattern could have been interpreted as a signed number and there is no protection against mixing such interpretations. So be careful! Similarly, the output is assigned an integer value that is automatically reinterpreted as 4 bits.

The fifth alternative is a default. At first glance, this seems redundant as 2 bits give four values, as specified. The control input, a, is of type logic, however. Therefore, its bits can take x or z values. So, in effect, there are 16 possible values for a. The default line assigns an x to the output if any of the input bits is not a true binary value. This line will not be synthesized, but it is good practice to include it if you want to check for unusual behavior in simulation.

4.2.2 Parameterizable Decoder

We have seen two ways to describe a 2 to 4 decoder. The same structures could easily be adapted to model a 3 to 8 decoder or a 4 to 16 decoder. Although these devices are clearly more complex than the 2 to 4 decoder, conceptually there is little difference. It would be convenient to have a general N to 2^N decoder that could be described once, but used for any application. We cannot, of course, write a case statement with an indeterminate number of branches. Another approach is needed. One way to think about this is based on the following observation. The output can be described as a single 1 shifted leftward by a number of places given by the input number. The bit pattern of the input, a, is again interpreted as an integer number.

```
y = 1'b1 << a;
```

We specify a single bit with the notation `1'b`, followed by the value of that bit.

Similarly, an array of size 2^N can be declared as `[(1<<N)-1:0]`. If N takes the default value of 3, the width of the output vector is given by 1 (note that this can be an integer value, not a bit value) shifted left by three places to give the bit pattern 1000_2, which is 8 in decimal. To get 8 bits in total, we make the range 7 down to 0.

We saw in the previous chapter that parameters can be used to pass values, such as delays, to SystemVerilog models. We can similarly use a parameter to define the size of a structure.

```
module decoderN #(parameter N = 3)
  (output logic [(1<<N)-1:0] y, input logic [N-1:0] a);

  always_comb
    y = 1'b1 << a;

endmodule
```

There is, of course, another way to describe the decoder—as a $\log_2(N)$ to N decoder. In fact, we need the *ceiling* of the function; in other words, the size of the result is rounded up to the next highest integer. This function, `clog2`, can be implemented by shifting and adding. In SystemVerilog, a *constant function* can be used to determine such values as array sizes. Constant functions are evaluated at compile time and hence are a little more limited than regular functions. An example of a constant function is given in the following. It is left as an exercise for the reader to understand how the `clog2` function works.

```
module decoderlogN #(parameter N = 8)
  (output logic [N-1:0] y,
   input logic [clog2(N)-1:0] a);

  function int clog2(input int n);
    begin
    clog2 = 0;
    n--;
    while (n > 0)
      begin
          clog2++;
          n >>= 1;
          end
    end
  endfunction

  always_comb
    y = 1'b1 << a;

endmodule
```

4.2.3 Seven-Segment Decoder

Sometimes, several input patterns might give the same output. There are two alternatives to the **case** statement that allow don't care values.

- **casez** allows z values in the case branches to be treated as don't cares. A ? can be used instead of z.
- **casex** allows z and x to be treated as don't cares.

If more than one pattern should give the same output, the patterns can be listed. For example, the following model describes a seven-segment decoder to display the digits 0 to 9. If the bit patterns corresponding to decimal values 10 to 15 are fed

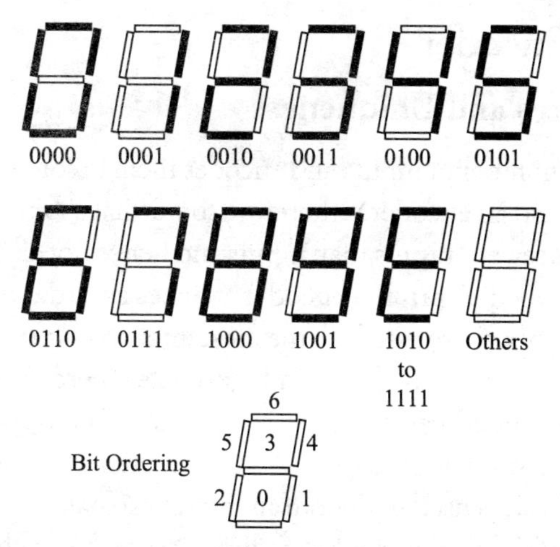

Figure 4.4 Seven-segment display.

into the decoder, an E (for "Error") is displayed. If the inputs contain Xs or other invalid values, the display is blanked. These patterns are shown in Figure 4.4. (But be careful, there are many different ways to encode seven-segment displays. This example will need to be changed if the segments are differently numbered or if the logic is active low.)

```
module sevenseg(output logic [6:0] data,
                input logic [3:0] address);

always_comb
  casez (address)
    4'b0000 : data = 7'b1110111;
    4'b0001 : data = 7'b0010010;
    4'b0010 : data = 7'b1011101;
    4'b0011 : data = 7'b1011011;
    4'b0100 : data = 7'b0111010;
    4'b0101 : data = 7'b1101011;
    4'b0110 : data = 7'b1101111;
    4'b0111 : data = 7'b1010010;
    4'b1000 : data = 7'b1111111;
    4'b1001 : data = 7'b1111011;
    4'b101?,
    4'b11?? : data = 7'b1101101;
    default : data = 7'b0000000;
  endcase
endmodule
```

4.3 Priority Encoder

4.3.1 Don't Cares and Uniqueness

An encoder takes a number of inputs and encodes them in some way. The difference between a decoder and an encoder is therefore somewhat arbitrary. In general, however, an encoder has fewer outputs than inputs and hence not all input combinations can be uniquely encoded. A priority encoder attaches an order of importance to the inputs. Thus, if two inputs are asserted, the most important input takes priority. The symbol for a priority encoder is shown in Figure 4.5. There are three outputs. The lower two are the encoded values of the four inputs. The upper output indicates whether the output combination is valid. An OR function (≥ 1) is used to check that at least one input is 1. Z is used to denote an internal signal. Thus, Z10 is connected to 10. This avoids unsightly and confusing lines across the symbol.

An example of a priority encoder is given in Table 4.2. The Valid output is used to signify whether at least one input has been asserted and hence whether the outputs are valid.

We can code this directly in SystemVerilog, using the **casez** statement. We need to be a little careful when using don't cares. It would be very easy to write two or more lines that overlapped. In other words, a pattern might match two or more case branches. For example, the pattern `4'b0110` would match both `4'b0?10` and `4'b01?0`. If both these alternatives were in a **casez** statement, the one occurring first would be selected in simulation. If the design were synthesized, however, there would be an ambiguity and the synthesis tool might attempt to impose its own priority. To avoid any ambiguity, it is good practice to qualify a **casez** statement with the **unique** modifier. If there is an overlap, an error would be flagged during compilation.

We can reproduce the structure of the truth table by making one assignment to y and valid simultaneously. Curly braces `{ }` are used to concatenate variables: in other words, to express two separate variables as one vector.

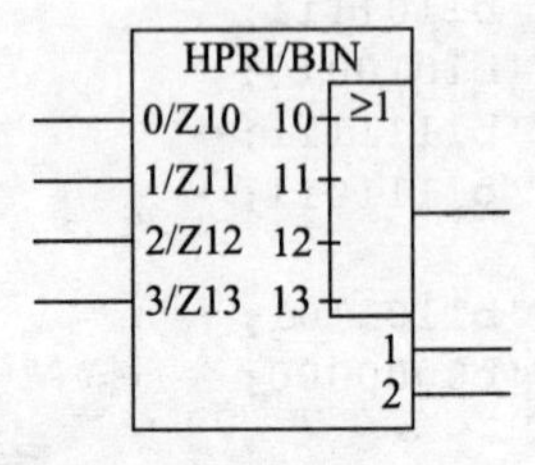

Figure 4.5 4 to 2 priority encoder.

Table 4.2 Priority Encoder

Inputs				Outputs		
A3	A2	A1	A0	Y1	Y0	Valid
0	0	0	0	0	0	0
0	0	0	1	0	0	1
0	0	1	-	0	1	1
0	1	-	-	1	0	1
1	-	-	-	1	1	1

```
module encoder (output logic [1:0] y, logic valid,
                input logic [3:0]a);

always_comb
  unique casez (a)
    4b'1??? : {y,valid} = 3'b111;
    4b'01?? : {y,valid} = 3'b101;
    4b'001? : {y,valid} = 3'b011;
    4b'0001 : {y,valid} = 3'b001;
    default : {y,valid} = 3'b000;
  endcase

endmodule
```

4.4 Adders

4.4.1 Functional Model

The IEEE symbol for a 4-bit adder is shown in Figure 4.6. The $\sum$ symbol denotes an adder. P and Q are assumed to be the inputs to the adder. CI and CO are carry in and carry out, respectively.

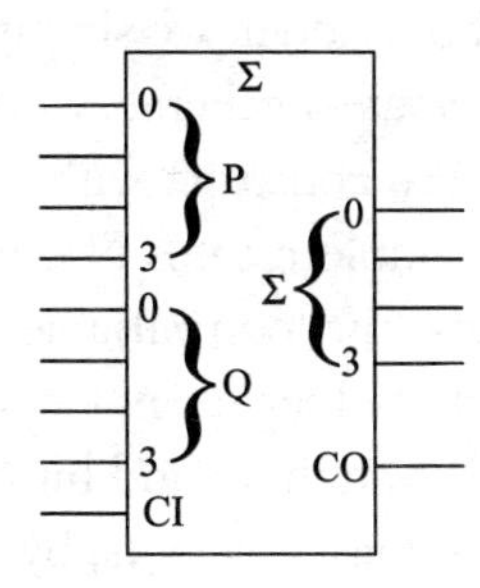

Figure 4.6 Four-bit adder.

The addition of two n-bit integers produces a result of length $n + 1$, where the most significant bit is the carry out bit. Therefore, within the SystemVerilog description we must separate the result into an n-bit sum and a carry out bit. The following code performs these actions for both signed and unsigned addition. The curly braces concatenate a single bit and an n-bit vector to give a vector of length $n + 1$. The complete code follows.

```
module adder #(parameter N = 4)
  (output logic [N-1:0] Sum, output logic Cout,
   input logic [N-1:0] A, B, input logic Cin);

  always_comb
    {Cout, Sum} = A + B + Cin;

endmodule
```

4.4.2 Ripple Adder

A simple model of a single-bit full adder (see Exercise 3.3) might be:

```
module fulladder (output logic sum, cout,
                  input logic a, b, cin);

  always_comb
    begin
    sum = a ^ b ^ cin;
    cout = a & b | a & cin | b & cin;
    end

endmodule
```

This model contains two assignments, sum and cout, written as Boolean expressions. We can build a multi-bit adder using several instances of this full adder. If we know how many bits will be in our adder, we simply instantiate the model several times. If, however, we want to create a general N-bit adder, we need some type of iterative construct. The **generate** construct with a **for** loop allows repetition in a dataflow description. This example creates N-2 instances and, through the Ca vector, wires them up. Notice that the loop variable, i, is declared as a **genvar**.

The first and last bits of the adder do not conform to the general pattern, however. Bit 0 should have Cin as an input and bit N-1 should generate Cout. We make special cases of the first and last elements, by instantiating them outside the generate block.

```
module ripple #(parameter N = 4)
  (output logic [N-1:0] Sum, output logic Cout,
   input logic [N-1:0] A, B, input logic Cin);

  logic [N-1:1] Ca;
  genvar i;

  fulladder f0 (Sum[0], Ca[1], A[0], B[0], Cin);

  generate for (i = 1; i < N-1; i++)
    begin : f_loop
    fulladder fi (Sum[i], Ca[i+1], A[i], B[i], Ca[i]);
    end
  endgenerate

  fulladder fN (Sum[N-1], Cout, A[N-1], B[N-1], Ca[N-1]);

endmodule
```

4.4.3 Tasks

The full adder could also be implemented as a **task**. A SystemVerilog **task** is a sub-routine, like a **function**, but without a return value. Tasks can include more diverse constructs than functions.

```
module ripple_task #(parameter N = 4)
  (output logic [N-1:0] Sum, output logic Cout,
   input logic [N-1:0] A, B, input logic Cin);

  logic [N-1:1] Ca;
  genvar i;

task automatic fulladder (output logic sum, cout,
                          input logic a, b, cin);

    begin
    sum = a ^ b ^ cin;
    cout = a & b | a & cin | b & cin;
    end

endtask

always_comb
  fulladder (Sum[0], Ca[1], A[0], B[0], Cin);
```

```
  generate for (i = 1; i < N-1; i++)
    begin : f_loop
    always_comb
      fulladder (Sum[i], Ca[i+1], A[i], B[i], Ca[i]);
    end
  endgenerate

always_comb
  fulladder fN (Sum[N-1], Cout, A[N-1], B[N-1], Ca[N-1]);

endmodule
```

The task is declared as **automatic** to ensure that each call has its own copy of variables. Otherwise, variables are shared between each call, which would lead to conflicts between assignments.

4.5 Parity Checker

The principle of parity checking was explained in Chapter 2. The IEEE symbol for a parity checker is shown in Figure 4.7. The symbol 2k indicates that the output is asserted if 2k inputs are asserted for any integer, k. Thus, the output is asserted for even parity. An odd parity checker has the output inverted.

This function can be implemented using a **for** loop. The syntax of this is the same as in the C programming language.

```
module parity_loop #(parameter N = 4)
       (output logic even, input logic [N-1:0] a);

  always_comb
    begin
    even = '1;
    for (int i = 0; i < N; i++)
      even = even ^ a[i];
    end

endmodule
```

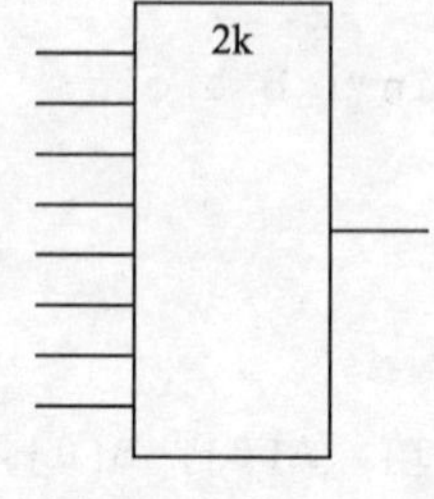

Figure 4.7 Even parity checker.

It is possible to do this in a much more concise way. In addition to the usual programming operators, SystemVerilog has reduction operators that can be applied to all the bits of a vector. For example, the even parity bit can be generated by taking the exclusive OR of all the bits of a vector and inverting.

```
module parity #(parameter N = 4)
      (output logic even, input logic [N-1:0] a);

  always_comb
    even = ~^a;

endmodule
```

4.6 Three-State Buffers

4.6.1 Multi-Valued Logic

In addition to the normal Boolean logic functions, it is possible to design digital hardware using switches to disconnect a signal from a wire. For instance, we can connect the outputs of several gates together, through switches, such that only one output is connected to the common wire at a time. This same functionality could be achieved using conventional logic, but would probably require a greater number of transistors. The IEEE symbol for a three-state buffer is shown in Figure 4.8. The symbol "1" shows the device is a buffer. "EN" is the symbol for an output enable and the inverted equilateral triangle indicates a three-state output.

If we write a model using variables of type **`logic`**, we must ensure that two models do not attempt to put a value onto the same variable. The purpose of using three-state buffers is to allow two or more component outputs to be connected together, provided that no more than one output generates a logic 1 or 0 and the rest of the outputs are in the high impedance state. This cannot be done with logic variables—a SystemVerilog simulator does not treat z as a special case. Resolution of conflicting logic values is done using a **`wire`**. Assignment of high-impedance can be done from within a procedural block, but it is easier to use a *continuous assignment* that is outside any procedural block. Conversely, most of the examples

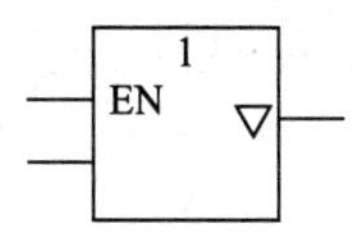

Figure 4.8 Three-state buffer.

in this chapter can be written using continuous assignments, but the procedural style is easier to use. Therefore, it is recommended that all three-state elements are modeled using continuous assignments and that the continuous assignment is only used for this purpose.

A SystemVerilog model of a three-state buffer follows.

```
module threestate (output wire y,
                   input logic a, enable);

  assign y = enable ? a : 'z;

endmodule
```

It is also possible to use three-state logic to build a multiplexer. A 4 to 1 multiplexer implemented in three-state logic follows. There are four assignments to y. At any time, three are z and one is an input value. In order for the output value to be correctly determined, and in order not to cause a compilation error, y must be declared to be a wire.

```
module threemux4 (output wire y,
                  input logic a, b, c, d, s0, s1);

  assign y = (~s0 && ~s1) ? a : 'z;
  assign y = (s0 && ~s1) ? b : 'z;
  assign y = (~s0 && s1) ? c : 'z;
  assign y = (s0 && s1) ? d :'z;

endmodule
```

4.7 Testbenches for Combinational Blocks

In the previous chapter, we introduced the idea of writing simulation testbenches in SystemVerilog for simple combinational circuits. Testbenches are not synthesizable and therefore the entire scope of SystemVerilog can be used to write them. Testbenches are also notable for the fact that their module declarations do not include any inputs or outputs—a testbench represents the rest of the world.

Two functions are generally performed in a testbench: generation of input stimuli and checking of results. The simple testbenches shown in the previous chapter did not perform any checking. Moreover, input stimuli were generated using concurrent assignments. This style is fine for simple circuits, but is not appropriate for circuits with multiple inputs. For example, let us write a testbench for the n-bit adder of Section 4.4.1.

```
module TestNBitAdder;

parameter N = 4;
logic Cin, Cout;
logic [N-1:0] Sum, A, B;

adder #(N) s0 (.*);

initial
  begin
  Cin = '0;
  A = 4'b0000;
  B = 4'b0000;
  #5ns A = 4'b1111;
  #5ns Cin = '1;
  #5ns A = 4'b0111;
  #5ns B = 4'b1111;
  #5ns Cin = '0;
  end

endmodule
```

The instantiation of the adder has a parameter (`N`) and uses a wild card `(.*)` to connect signals. This is allowable if the wire and variable names in the testbench are *exactly* the same as those in the module being instantiated.

Note that the time is relative (we wait for 5 ns at a time), rather than absolute. Remember, too, that **`initial`** indicates a procedure that is executed once, not an initialization of variables.

As far as combinational circuits are concerned, this is about as complex as we ever need to get. It is difficult, however, to work out what is going on. For example, we try to add "0111" to "0000" with a carry in bit of 1. The simulation tells us that the sum is "1000" with a carry out bit of 0. It is just about possible to work out that this is correct, but it is not easy. Instead, we could use integers.

```
initial
  begin
  Cin = 0;
  A = 0;
  B = 0;
  #5ns A = 15;
  #5ns Cin = 1;
  #5ns A = 7;
  #5ns B = 15;
  #5ns Cin = 0;
  end
```

Now we can see that 7+0+1 is equal to 8 (with no carry out). Better still, we could let the testbench itself check the addition. In general, we do not necessarily want to be told that the design is correct, but we do want to know if there is an error. In Chapter 5, we will see how warning messages can be generated. Another technique is to generate an error signal when unexpected behavior occurs. It is then relatively easy to spot one signal changing state in a long simulation with a lot of complex data. To the testbench above, we simply add an error signal:

```
logic error;
```

together with a process that is triggered whenever one of the outputs from the adder changes:

```
always @(Cout, Sum)
  error = ((A + B + Cin) != Sum);
```

The idea is to check the operation by performing it in a different way. In later chapters, we will again see this principle. We will also use processes, triggered by changing signals, to monitor outputs.

Summary

In this chapter, we introduced a number of typical combinational building blocks. The IEEE standard symbols for these blocks were described. We briefly introduced testbenches for combinational logic blocks.

Further Reading

A full description of the IEEE symbols is given in the IEEE standard and in a number of digital design textbooks. Manufacturers' data sheets may use the IEEE symbols or a less standard form.

Exercises

4.1 SystemVerilog models can be written using continuous and procedural assignments. Explain, with examples, the meaning of continuous and procedural in this context.

4.2 Write a SystemVerilog model for the function $Z = A \cdot B + C \cdot D$.

4.3 Write SystemVerilog models of a 3 to 8 decoder using (a) Boolean operators, (b) a conditional operator, and (c) a shift operator. Write a testbench to compare the three versions.

4.4 Write a SystemVerilog model of a 2^n to n priority encoder.

4.5 Write a model of an n-input multiplexer. Write a suitable testbench.

4.6 A comparator is used to to determine whether two signals have equal values. A one-bit comparator is described by

```
eqo = ~(x ^ y) & eqi;
```

where `eqi` is the result of the comparison of other bits and `eqo` is passed to the next comparison operation. Write a model of an n-bit iterative comparator.

5 SystemVerilog Models of Sequential Logic Blocks

In the previous chapter we presented several examples of combinational building blocks, at the same time introducing various aspects of SystemVerilog. In this chapter we shall repeat the exercise for sequential blocks.

5.1 Latches

5.1.1 SR Latch

There is often confusion between the terms *latch* and *flip-flop*. Here, we will use *latch* to mean a level-sensitive memory device and *flip-flop* to specify an edge-triggered memory element. We discuss the design of latches and flip-flops in Chapter 13. We simply note here that a latch is based on cross-coupled gates, as shown in Figure 5.1. Table 5.1 gives the truth table of this latch.

When S and R are both at logic 1, the latch holds onto its previous value. When both are at 0, both outputs are at 1. It is this latter behavior that makes the SR latch unsuitable for designing larger circuits, as a latch or flip-flop would normally be expected to have different values at its two outputs, and it is difficult to ensure that both inputs will never be 0 at the same time.

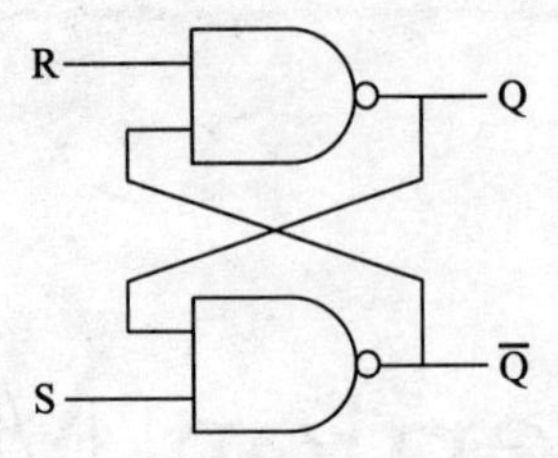

Figure 5.1 SR latch.

The SR latch could be modeled in SystemVerilog in a number of ways. Two examples follow.

```
module rslatch1 (output wire q, qbar,
                 input logic r, s);

  nand n0(q, qbar, r);
  nand n1(qbar, q, s);

endmodule
```

In the first example, the latch is modeled using two NAND gates. There is nothing fundamentally wrong with this model, but it is dependent on the technology, and it would be a little impractical for larger elements.

```
module rslatch2 (output logic q, qbar,
                 input logic r, s);

always @(r, s)
  unique case ({r, s})
    2'b00: {q, qbar} <= 2'b11;
    2'b01: {q, qbar} <= 2'b10;
    2'b10: {q, qbar} <= 2'b01;
    default;
  endcase

endmodule
```

In the second model, we explicitly model an element with storage. Therefore, we cannot use an **always_comb** procedural block, which would imply purely

Table 5.1 Truth Table of SR Latch

S	R	Q	$\bar{Q}$
0	0	1	1
0	1	0	1
1	0	1	0
1	1	Q	$\bar{Q}$

combinatorial logic, without storage. We can use a general purpose **always** block, as shown. We have to list the two inputs and, using a **case** statement, the truth table of the latch can be reproduced. The curly braces { } concatenate two variables, both in the case selector and in the case branches.

In the first three branches of the **case** statement, values are assigned to q and qbar depending on the combination of inputs. Nothing is assigned in the fourth, **default** branch, so the q and qbar values are retained. In other words, the values are latched. If the module is synthesized, a latch will be inferred.

Notice that we have specified that the case statement is non-overlapping (**unique**) and that there is a default (in which nothing happens).

These two examples show that omitting an assignment for one or more input conditions infers a latch. If this is done by accident in an **always_comb** block, a synthesis tool will generate a warning *but* might interpret the code as a latch. Such warnings should always be examined carefully. Unintended latches will almost certainly cause the circuit to work incorrectly.

5.1.2 D Latch

Because an SR latch can have both outputs at the same value, it is seldom if ever used. More useful is the D latch, as shown in Figure 5.2. The input of a D latch is transferred to the output if an enable signal is asserted. 1D indicates a dependency of the D input on control signal 1 (C1). The $\bar{Q}$ output is not shown.

A behavioral SystemVerilog model of a D latch is

```
module dlatch (output logic q, input logic d, en);

always_latch
  if (en)
    q <= d;

endmodule
```

The assignment is *nonblocking*, as shown by the symbol <=. Nonblocking assignments are completed after blocking assignments (=). Sequential logic should always be modeled with nonblocking assignments to ensure correct simulation behavior.

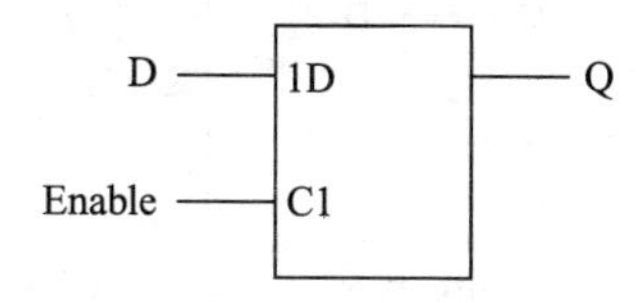

Figure 5.2 Level-sensitive D latch.

5.2 Flip-Flops

5.2.1 Edge-Triggered D Flip-Flop

In the next chapter, the principle of synchronous sequential design is described. The main advantage of this approach to sequential design is that all changes of state occur at a clock *edge*. The clock edge is extremely short in comparison to the clock period and to propagation delays through combinational logic. In effect, a clock edge can be considered to be instantaneous.

The IEEE symbol for a positive edge-triggered D flip-flop is shown in Figure 5.3. Again, the number 1 shows the dependency of D on C. The triangle at the clock input denotes edge-sensitive behavior. An inversion circle, or its absence, shows sensitivity to a negative or positive edge, respectively.

The simplest SystemVerilog model of a positive edge-triggered D flip-flop is given in the following.

```
module dff (output logic q, input logic d, clk);

always_ff @(posedge clk)
  q <= d;

endmodule
```

Again, a *nonblocking* assignment is used, as this is sequential logic. Similarly, a negative edge-triggered flip-flop can be modeled by detecting a transition to logic 0.

5.2.2 Asynchronous Set and Reset

When power is first applied to a flip-flop, its initial state is unpredictable. In many applications, this is unacceptable, so flip-flops are provided with further inputs to set (or reset) their outputs to 1 or to 0, as shown in Figure 5.4. Notice that the absence of any dependency on the clock implies asynchronous behavior for R and S.

These inputs should only be used to initialize a flip-flop. It is very bad practice to use these inputs to set the state of a flip-flop during normal system operation. The reason for this is that in synchronous systems, flip-flops only change state when

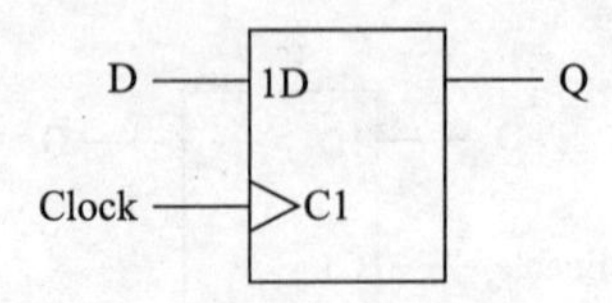

Figure 5.3 Positive edge-triggered D flip-flop.

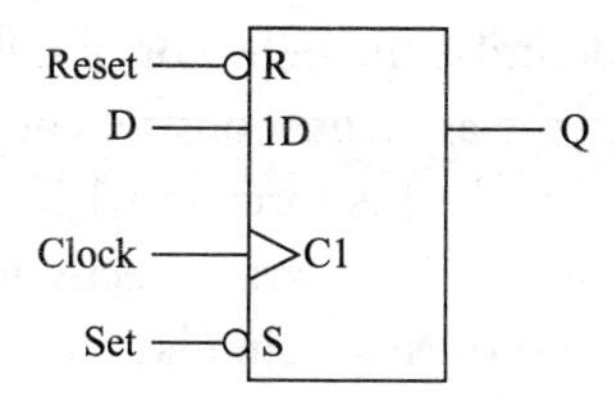

Figure 5.4 Positive edge-triggered D flip-flop with asynchronous reset and set.

clocked. The set and reset inputs are *asynchronous* and hence cannot be guaranteed to change an output at a particular time. This can lead to all sorts of timing problems. In general, keep all designs strictly synchronous or follow a structured asynchronous design methodology.

A SystemVerilog model of a flip-flop with an asynchronous reset must respond to changes in the clock and in the reset input.

```
module dffr (output logic q,
             input logic d, clk, n_reset);

always_ff @(posedge clk, negedge n_reset)
  if (~n_reset)
    q <= '0;
  else
    q <= d;

endmodule
```

An asynchronous set can be described in a similar way (see Exercises).

It is possible for a flip-flop to have both an asynchronous set and reset. For example:

```
module dffrs (output logic q,
              input logic d, clk, n_reset, n_set);

always_ff @(posedge clk, negedge n_reset,
            negedge n_set)
  if (~n_set)
    q <= '1;
  else if (~n_reset)
    q <= '0;
  else
    q <= d;

endmodule
```

This may not correctly describe the behavior of a flip-flop with asynchronous inputs because asserting both the asynchronous set and reset is usually considered an illegal operation. In this model, Q is forced to 1 if n_set is 0, regardless of the n_reset signal. Even if this model synthesizes correctly, we would still wish to check that this condition did not occur during a simulation.

5.2.3 Synchronous Set and Reset and Clock Enable

Flip-flops may have synchronous set and reset functions as well as, or instead of, asynchronous set or reset inputs. A synchronous set or reset only takes effect at a clock edge. Thus, a SystemVerilog model of such a function must include a check on the set or reset input after the clock edge has been checked. It is not necessary to include synchronous set or reset inputs in the excitation list because the process is only activated at a clock edge. This is shown in IEEE notation in Figure 5.5. *R* is now shown to be dependent on *C* and is therefore synchronous.

```
module dffsr (output logic q,
              input logic d, clk, n_reset);

always_ff @(posedge clk)
  if (~n_reset)
    q <= '0;
  else
    q <= d;

endmodule
```

Notice that the only difference between the synchronous and asynchronous reset is whether the signal appears in the excitation list of the **always_ff** block.

Similarly, a flip-flop with a clock enable signal may be modeled with that signal checked after the edge detection. In Figure 5.6, the dependency notation shows that *C* is dependent on *G* and *D* is dependent on (the edge-triggered behavior of) *C*.

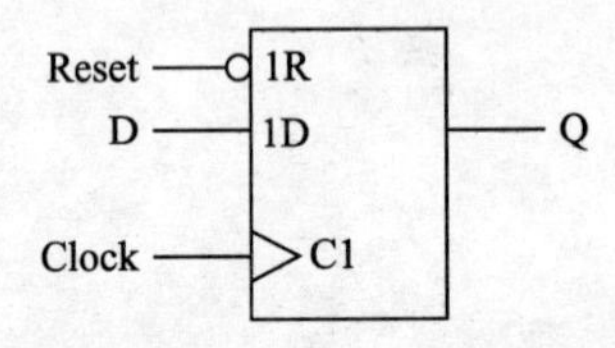

Figure 5.5 Positive edge-triggered D flip-flop with synchronous reset.

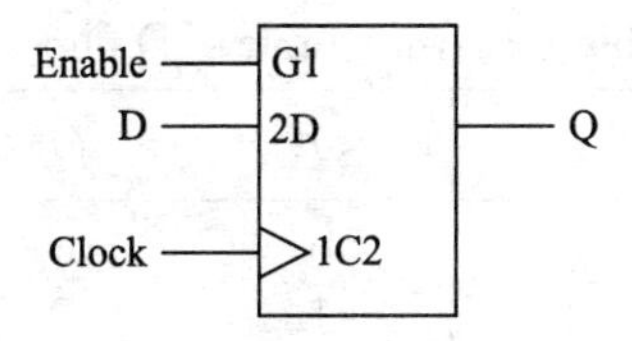

Figure 5.6 Positive edge-triggered D flip-flop with clock enable.

```
module dffe (output logic q,
             input logic d, clk, enable);

always_ff @(posedge clk)
  if (enable)
    q <= d;

endmodule
```

A synthesis system is likely to interpret this as a flip-flop with a clock enable. The following model is likely to be interpreted differently, although it appears to have the same functionality.

```
module dffce (output logic q,
              input logic d, clk, enable);

logic ce;

always_comb
  ce = enable & clk;

always_ff @(posedge ce)
  q <= d;

endmodule
```

Again, the *D* input is latched if enable is true and there is a clock edge. This time, however, the clock signal passes through an AND gate and hence is delayed. The *D* input is also latched if the clock is true and there is a rising edge on the enable signal. This is another example of design that is not truly synchronous and that is therefore liable to timing problems. This style of design should generally be avoided, although for low-power applications, the ability to turn off the clock inputs to flip-flops can be useful.

Table 5.2 Truth Table of D Flip-Flop

D	Q^+	$\bar{Q}^+$
0	0	1
1	1	0

5.3 JK and T Flip-Flops

A D flip-flop registers its input at a clock edge, making that value available during the next clock cycle. JK and T flip-flops change their output states at the clock edge in response to their inputs and to their present states. Truth tables for D, JK, and T flip-flops are in Tables 5.2, 5.3, and 5.4, respectively.

Both the Q and $\bar{Q}$ outputs are shown. Symbols for D, JK, and T flip-flops with both outputs and with a reset are shown in Figure 5.7.

```
module jkffr (output logic q, qbar,
              input logic j, k, clk, n_reset);

always_ff @(posedge clk, negedge n_reset)
  if (~n_reset)
    {q, qbar} <= {1'b0, 1'b1};
  else
    case ({j, k})
      2'b11 : {q, qbar} <= {qbar, q};
      2'b10 : {q, qbar} <= {1'b1, 1'b0};
```

Table 5.3 Truth Table of JK Flip-Flop

J	K	Q^+	$\bar{Q}^+$
0	0	Q	$\bar{Q}$
0	1	0	1
1	0	1	0
1	1	$\bar{Q}$	Q

Table 5.4 Truth Table of T Flip-Flop

T	Q^+	$\bar{Q}^+$
0	Q	$\bar{Q}$
1	$\bar{Q}$	Q

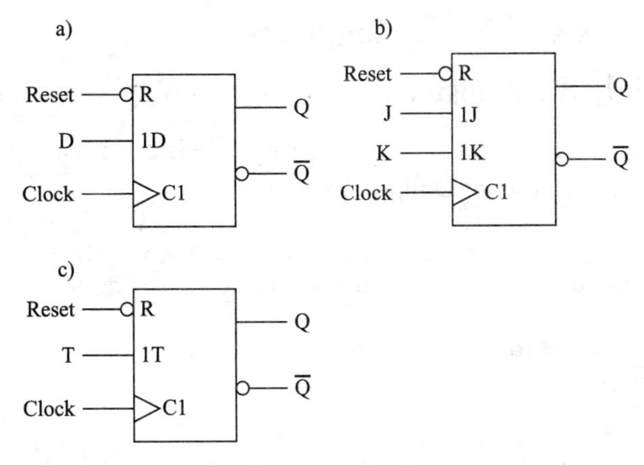

Figure 5.7 (a) D flip-flop; (b) JK flip-flop; (c) T flip-flop.

```
      2'b01 : {q, qbar} <= {1'b0, 1'b1};
      default:;
    endcase

endmodule
```

A case statement determines the internal state of the JK flip-flop. The selector of the case statement is formed by concatenating the *J* and *K* inputs. The **default** clause covers the 00 case and other undefined values. Nothing is done in that clause, so the internal state is retained.

```
module tffr (output logic q, qbar,
             input logic t, clk, n_reset);

always_ff @(posedge clk, negedge n_reset)
  if (~n_reset)
    {q, qbar} <= {1'b0, 1'b1};
  else
    if (t)
      {q, qbar} <= {qbar, q};

endmodule
```

The internal state of the T flip-flop is retained between activations of the procedural block, if the T input is not set.

5.4 Registers and Shift Registers

5.4.1 Multiple Bit Register

A D flip-flop is a 1-bit register. Thus, if we want a register with more than 1 bit, we simply need to define a set of D flip-flops using vectors:

```
module dffn #(parameter N = 8) (output logic [N-1:0]q,
    input logic [N-1:0] d, input logic clk, n_reset);

always_ff @(posedge clk, negedge n_reset)
  if (~n_reset)
    q <= '0;
  else
    q <= d;

endmodule
```

The IEEE symbol for a 4-bit register is shown in Figure 5.8. Note that the common signals are contained in a control block, drawn as a rectangle with the lower corners cut off.

5.4.2 Shift Registers

An extension of the previous model of a register includes the ability to shift the bits of the register to the left or to the right. For example, a sequence of bits can be converted into a word by shifting the bits into a register, and moving the bits along at each clock edge. After a sufficient number of clock edges, the bits of the word are available as a single word. This is known as a *serial-in, parallel-out* (SIPO) register.

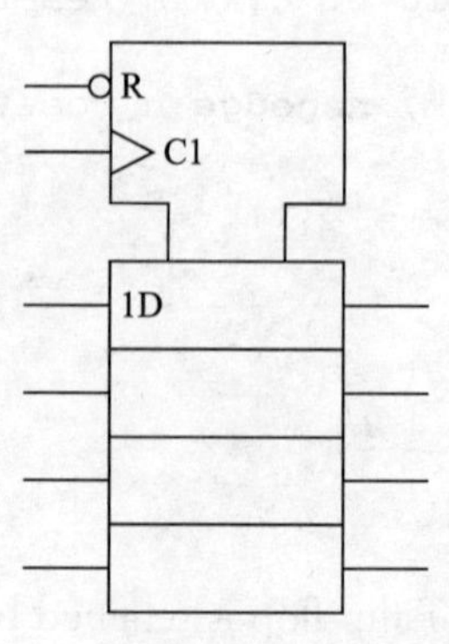

Figure 5.8 Four-bit register.

Table 5.5 Universal Shift Register

$S_1 S_0$	Action
00	Hold
01	Shift right
10	Shift left
11	Parallel load

```
module sipo #(parameter N = 8) (output logic [N-1:0] q,
                   input logic a, clk);

  always_ff @(posedge clk)
    q <= {q[N-2:0], a};

endmodule
```

At each clock edge, the bits of the register are moved along by 1, and the input, a, is shifted into the 0th bit. The assignment does this by assigning bits $n-2$ to 0 to bits $n-1$ to 1, respectively, and concatenating a to the end of the assignment. The old value for bit $n-1$ is lost.

A more general shift register is the universal shift register, Table 5.5. This can shift bits to the left or to the right, and can load an entire new word in parallel. To do this, two control bits are needed. The IEEE symbol is shown in Figure 5.9.

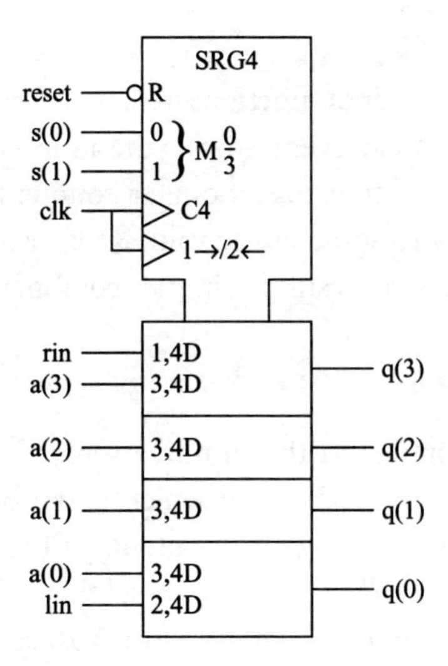

Figure 5.9 Universal shift register.

There are four control modes shown by $M\frac{0}{3}$. The clock signal is split into two for convenience. Control signal 4 is generated, and in modes 1 and 2, a shift left or shift right operation, respectively, is performed. 1,4*D* means that a D-type operation occurs in mode 1 when control signal 4 is asserted.

```
module usr #(parameter N = 8) (output logic [N-1:0]q,
  input logic [N-1:0] a, input logic [1:0] s,
  input logic lin, rin, clk, n_reset);

always_ff @(posedge clk, negedge n_reset)
  if (~n_reset)
    q <= '0;
  else
    case (s)
      2'b11: q <= a;
      2'b10: q <= {q[n-2:0], lin};
      2'b01: q <= {rin, q[n-1:1]};
      default:;
    endcase
endmodule
```

The shift operations are done by taking the lowest $(n-1)$ bits and concatenating the leftmost input (shift left) or by taking the upper $(n-1)$ bits concatenated to the rightmost input (shift right). It would be possible to use the shift operators, but in practice they are not needed.

5.5 Counters

Counters are used for a number of functions in digital design, for example, counting the number of occurrences of an event, storing the address of the current instruction in a program, or generating test data. Although a counter typically starts at zero and increments monotonically to some larger value, it is also possible to use different sequences of values, which can result in simpler combinational logic.

5.5.1 Binary Counter

A binary counter is a counter in the intuitive sense. It consists of a register of a number of D flip-flops, the content of which is the binary representation of a decimal number. At each clock edge the contents of the counter are increased by one, as shown in Figure 5.10. We can easily model this in SystemVerilog using the + operator. The reset operation is shown in Figure 5.10 as setting the contents (*CT*) to 0. The weight of each stage is shown in brackets.

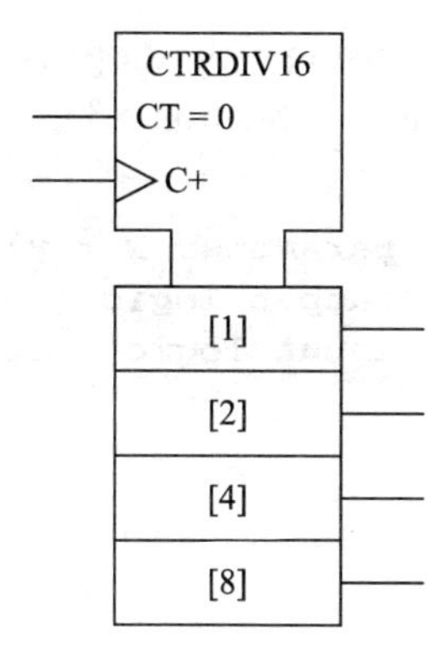

Figure 5.10 Binary counter.

```
module bincounter #(parameter N = 8)
                  (output logic [N-1:0] count,
                   input logic n_reset, clk);

always_ff @(posedge clk, negedge n_reset)
  if (~n_reset)
    count <= 0;
  else
    count <= count + 1;

endmodule
```

Note that the + operator does not generate a carry out. Thus, when the counter has reached its maximum integer value (all 1s), the next clock edge will cause the counter to "wrap round," and its next value will be zero (all 0s). We could modify the counter to generate a carry out, but in general counters are usually designed to detect the all-1s state and to output a signal when that state is reached. A carry out signal would be generated one clock cycle later. It is trivial to modify this counter to count down, or to count by a value other than one (possibly defined by a parameter—see the exercises at the end of this chapter). It would be incorrect to use the increment operator "++", for example, by writing `count++;` instead of the assignment. Although more concise, the increment is a blocking assignment (equivalent to `count = count + 1;`). Using blocking assignments in sequential logic can cause erroneous simulated behavior.

The advantage of describing a counter in SystemVerilog is that the underlying combinational next state logic is hidden. For a counter with eight or more bits, the combinational logic can be very complex, but a synthesis system will generate that logic automatically. A simpler form of binary counter is the ripple counter.

An example of a ripple counter using T flip-flops is described in SystemVerilog in the following, using the T flip-flop of Section 5.3.

```
module ripple_counter #(parameter N = 8)
                       (output logic [N-1:0] count,
                        input logic n_reset, clk);

  logic [N:1] Ca;
  genvar i;

  tffr t0 (count[0], Ca[1], '1, clk, n_reset);

  generate for (i = 1; i < N; i++)
    begin : t_loop
    tffr ti (count[i], Ca[i+1], '1, Ca[i], n_reset);
    end
  endgenerate

endmodule
```

Note that the *T* input is held at a constant value in the description. When simulated using the T flip-flop model, this circuit behaves identically to the RTL model.

The ripple counter is, however, asynchronous. The second flip-flop is clocked from the Q output of the first flip-flop, as shown in Figure 5.11. A change in this output is delayed relative to the clock. Hence, the second flip-flop is clocked by a signal behind the true clock. With further counter stages, the delay is increased. Further, incorrect intermediate values are generated. Provided the clock speed is sufficiently slow, a ripple counter can be used instead of a synchronous counter, but in many applications a synchronous counter is preferred.

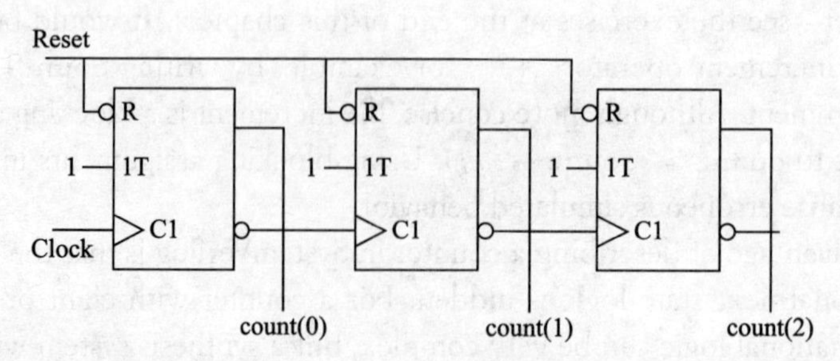

Figure 5.11 Ripple counter.

5.5.2 Johnson Counter

A Johnson counter (also known as a Möbius counter, after a Möbius strip, which is a strip of paper formed into a circle with a single twist, resulting in a single surface) is built from a shift register with the least significant bit inverted and fed back to the most significant bit, as shown in Figure 5.12.

An n-bit binary counter has 2^n states. An n-bit Johnson counter has $2n$ states. The advantage of a Johnson counter is that it is simple to build (like a ripple counter), but is synchronous. The disadvantage is the large number of unused states that form an autonomous counter in their own right. In other words, we have the intended counter and a *parasitic* state machine coexisting in the same hardware. Normally, we should be unaware of the parasitic state machine, but if the system somehow entered one of the unused states, the subsequent behavior might be unexpected. A SystemVerilog description of a Johnson counter follows.

```
module johnson #(parameter N = 8)
               (output logic [N-1:0] q,
                input logic clk, n_reset);

always_ff @(posedge clk, negedge n_reset)
  if (~n_reset)
    q <= '0;
  else
    q <= {~q[0], q[N-1:1]};

endmodule
```

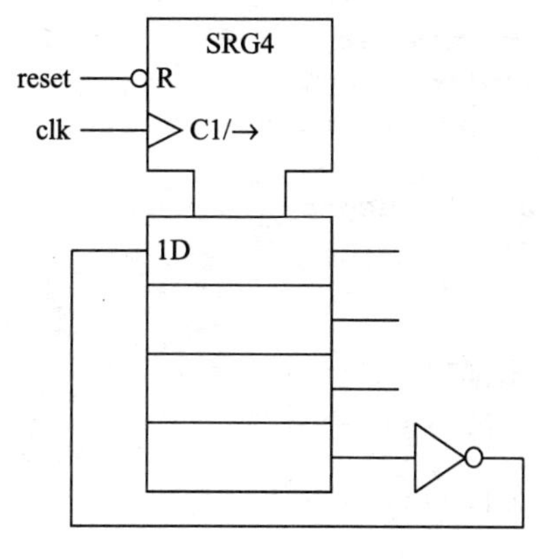

Figure 5.12 Johnson counter.

Table 5.6 Johnson Counter

Normal Counting Sequence	Parasitic Counting Sequence
0000	0010
1000	1001
1100	0100
1110	1010
1111	1101
0111	0110
0011	1011
0001	0101

The counting sequence of a 4-bit counter, together with the sequence belonging to the parasitic state machine, is shown in Table 5.6. Whatever the size of n, the unused states form a single parasitic counter with $2^n - 2n$ states.

Both sequences repeat but do not intersect at any point. The parasitic set of states of a Johnson counter should never occur, but if one of the states did occur somehow, perhaps because of a power supply glitch or because of some asynchronous input, the system can never return to its normal sequence. One solution to this is to make the counter *self-correcting*. It would be possible to detect every one of the parasitic states and to force a synchronous reset, but for an n-bit counter that is difficult. An easier solution is to note that the only legal state with a 0 in both the most significant and least significant bits is the all zeros state. On the other hand, three of the parasitic states have zeros in those positions. Provided that we are happy to accept that if the system does enter an illegal state it does not have to correct itself immediately, but can re-enter the normal counting sequence after a few clock cycles, we can simply detect any states that have a 0 at the most and least significant bits and force the next state to be 1000 or its n-bit equivalent.

```
module scjohnson #(parameter N = 8)
                  (output logic [N-1:0] q,
                   input logic clk, n_reset);

always_ff @(posedge clk, negedge n_reset)
  if (~n_reset)
    q <= '0;
  else
    if (~q[N-1] & ~q[0])
      q <= {1'b1, {(N-1){1'b0}}};
    else
      q <= {~q[0], q[N-1:1]};

endmodule
```

5.5.3 Linear Feedback Shift Register

Another counter that is simple in terms of next state logic is the linear feedback shift register (LFSR). This has $2^n - 1$ states in its normal counting sequence. The sequence of states appears to be random, hence the other name for the register: pseudo-random sequence generator (PRSG). The next state logic is formed by exclusive OR gates as shown in Figure 5.13.

There are a large number of possible feedback connections for each value of n that give the maximal length $(2^n - 1)$ sequence, but it can be shown that no more than four feedback connections (and hence three exclusive OR gates) are ever needed. The single state missing from the sequence is the all-0s state. Hence, the asynchronous initialization should be a "set." As with the Johnson counter, the LFSR could be made self-correcting. A SystemVerilog model of an LFSR valid for certain values of n is shown in the following.

The main advantage of using an LFSR as a counter is that nearly the full range of possible states $(2^n - 1)$ can be generated using simple next state logic. Moreover, the pseudo-random sequence can be exploited for applications such as coding.

In the SystemVerilog model, the feedback connections for LFSRs with 1 to 36 stages are defined in an **`initial`** block. Note that the model is only defined for the range 1 to 36 (with a default value of 8). Any attempt to use this model for a larger LFSR would result in an invalid model. The model defines up to three feedback connections—it is assumed that bit 0 is always used. The positions corresponding to the feedback connections are set to 1, using a shift operator and OR-ing theses values with the initial, all 0s value, using the assign and OR operator, |=. The **`initial`** block is evaluated once, at elaboration time, and the resulting value is used to configure the model.

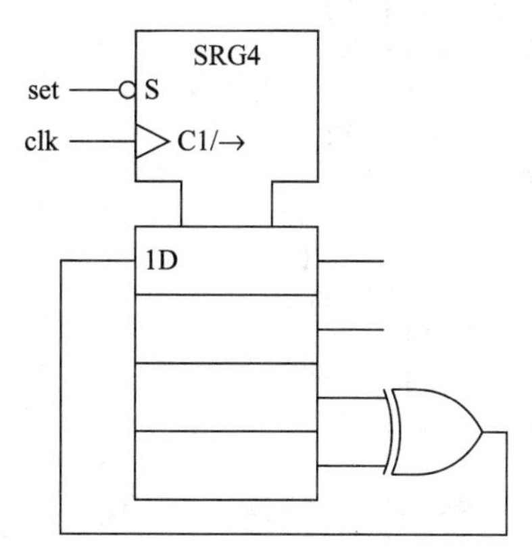

Figure 5.13 LFSR.

To construct the feedback connection for a particular size of LFSR, the stages of the LFSR referenced in the taps vector are XORed together using a **for** loop.

```
module lfsr #(parameter N = 4)
             (output logic [N-1:0] q,
              input logic clock, n_set);

  logic feedback;
  int i;
  logic [N-1:0] taps;

initial
begin
  taps = '0;
  case (N)
   2: taps |= (1'b1 << 1);
   3: taps |= (1'b1 << 1);
   4: taps |= (1'b1 << 1);
   5: taps |= (1'b1 << 2);
   6: taps |= (1'b1 << 1);
   7: taps |= (1'b1 << 1);
   8: taps |= ((1'b1 << 6) | (1'b1 << 5)
              | (1'b1 << 1));
   9: taps |= (1'b1 << 4);
  10: taps |= (1'b1 << 3);
  11: taps |= (1'b1 << 2);
  12: taps |= ((1'b1 << 7) | (1'b1 << 4)
              | (1'b1 << 3));
  13: taps |= ((1'b1 << 4) | (1'b1 << 3)
              | (1'b1 << 1));
  14: taps |= ((1'b1 << 12) | (1'b1 << 11)
              | (1'b1 << 1));
  15: taps |= (1'b1 << 1);
  16: taps |= ((1'b1 << 5) | (1'b1 << 3)
              | (1'b1 << 2));
  17: taps |= ((1'b1 << 3));
  18: taps |= ((1'b1 << 7));
  19: taps |= ((1'b1 << 6) | (1'b1 << 5)
              | (1'b1 << 1));
  20: taps |= (1'b1 << 3);
  21: taps |= (1'b1 << 2);
  22: taps |= (1'b1 << 1);
  23: taps |= (1'b1 << 5);
  24: taps |= ((1'b1 << 4) | (1'b1 << 3)
              | (1'b1 << 1));
  25: taps |= (1'b1 << 3);
```

```
  26: taps |= ((1'b1 << 8) | (1'b1 << 7)
               | (1'b1 << 1));
  27: taps |= ((1'b1 << 8) | (1'b1 << 7)
               | (1'b1 << 1));
  28: taps |= (1'b1 << 3);
  29: taps |= (1'b1 << 2);
  30: taps |= ((1'b1 << 16) | (1'b1 << 15)
               | (1'b1 << 1));
  31: taps |= (1'b1 << 3);
  32: taps |= ((1'b1 << 28) | (1'b1 << 27)
               | (1'b1 << 1));
  33: taps |= (1'b1 << 13);
  34: taps |= ((1'b1 << 15) | (1'b1 << 14)
               | (1'b1 << 1));
  35: taps |= (1'b1 << 2);
  36: taps |= (1'b1 << 11);
  endcase
end

always_ff @(posedge clock, negedge n_set)
  if (~n_set)
    q <= '1;
  else
    q <= {feedback, q[N-1:1]};

always_comb
  begin
  feedback = q[0];
  for (i = 1; i <= N - 1; i++)
    if (taps[i])
      feedback ^= q[i];
  end

endmodule
```

5.6 Memory

Computer memory is often classified as ROM (read-only memory) and RAM (random access memory). These are to some extent misnomers—ROM is random access and RAM is better thought of as read and write memory. RAM can further be divided into SRAM (static RAM) and DRAM (dynamic RAM). SRAM retains its contents while power is applied to the system. DRAM uses capacitors to store bits, which means that the capacitance charge can leak away with time. Hence, DRAM needs refreshing intermittently.

5.6.1 ROM

The contents of a ROM chip are defined once. Hence, we can use a constant array to model a ROM device in SystemVerilog. The seven-segment decoder from Chapter 4 described as a ROM is shown in the following.

```
module sevensegrom(output logic [6:0] data,
                   input logic [3:0] address);

  logic [6:0] rom [0:15] = {7'b1110111, //0
                             7'b0010010, //1
                             7'b1011101, //2
                             7'b1011011, //3
                             7'b0111010, //4
                             7'b1101011, //5
                             7'b1101111, //6
                             7'b1010010, //7
                             7'b1111111, //8
                             7'b1111011, //9
                             7'b1101101, //E 10
                             7'b1101101, //E 11
                             7'b1101101, //E 12
                             7'b1101101, //E 13
                             7'b1101101, //E 14
                             7'b1101101}; //E 15

always_comb
  data = rom[address];
endmodule
```

The variable `rom` is declared as a vector of 16 values that is, in turn, comprised of 7-bit vectors. Because no values can be written into the ROM, we can think of the device as combinational logic. In general, combinational logic functions can be implemented directly in ROM. Programmable forms of ROM are available (EPROM—electrically programmable ROM), but such devices require the application of a large negative voltage (–12 V) to a particular pin of the device. Such functionality is not modeled, as it does not form part of the normal operating conditions of the device.

5.6.2 SRAM

SRAM may be modeled in much the same way as ROM. Because data may be stored in the RAM as well as read from it, the `data` signal is declared to be a port of type **inout**, and because it can be put into a high-impedance state, it is declared as a

wire. In addition, three control signals are provided. The first, CS (chip select) is a general control signal to enable a particular RAM chip. The address range, in this example, is 0 to 15. If we were to use, say, four identical chips to provide RAM with an address range of 0 to 63 (6 bits), the upper two address bits would be decoded such that at any one time exactly one of the RAM chips is enabled by its CS signal. Hence, if the CS signal is not enabled, the data output of the RAM chip should be in the high-impedance state. The other two signals are OE (output enable) and WE (write enable). Only one of these two signals should be asserted at one time. Data is either read from the RAM chip when the OE signal is asserted, or written to the chip if the WE signal is asserted. If neither signal is asserted, the output remains in the high-impedance state. All the control signals are active low.

Like in the ROM, the memory array is modeled as a vector of vectors.

```
module RAM16x8(inout wire [7:0] Data,
               input logic [3:0]Address,
               input logic n_CS, n_WE, n_OE);

  logic [7:0] mem [0:15];

  assign Data = (~n_CS & ~n_OE) ? mem[Address] : 'z;

  always_latch
    if (~n_CS & ~n_WE & n_OE)
      mem[Address] <= Data;

endmodule
```

5.6.3 Synchronous RAM

The SRAM model is asynchronous and intended for modeling separate memory chips. Sometimes we wish to allocate part of an FPGA as RAM. In order for this to be synthesized correctly and for ease of use, it is best to make this RAM synchronous. Depending on the technology, there may be a variety of possible RAM structures, for example, synchronous read, dual-port. Here, we will simply show how a basic synchronous RAM is modeled. This parameterizable example can be synthesized in most programmable technologies.

```
module SyncRAM #(parameter M = 4, N = 8)
  (output logic [N-1:0] Qout,
   input logic [M-1:0] Address,
```

```
    input logic [N-1:0]Data, input logic WE, Clk);

  logic [N-1:0] mem [0:(1 << M)-1];

always_comb
  Qout = mem[Address];

always_ff @(posedge Clk)
  if (~WE)
    mem[Address] <= Data;

endmodule
```

The structure of the write part of this code is almost identical to that of a flip-flop with an enable—in this case, the enable signal is the WE input. As with the SRAM example, the Address input is interpreted as an unsigned integer to reference an array. This example could be extended to include an output enable and chip select, as above.

5.7 Sequential Multiplier

Let us consider a multiplier for two's complement binary numbers. Multiplication, whether decimal or binary, can be broken down into a sequence of additions. A SystemVerilog statement such as

```
  q = a * b;
```

where a and b are n-bit representations of (positive) integers, would be interpreted by a SystemVerilog synthesis tool as a combinational multiplication requiring n^2 full adders. If a and b are two's complement numbers, there also needs to be a sign adjustment. A combinational multiplier would take up a significant percentage of an FPGA for $n = 16$.

The classic trade-off in digital design is between area and speed. In this case, we can significantly reduce the area required for a multiplier if the multiplication is performed over several clock cycles. Between additions, one of the operands of a multiplication operation has to be shifted. Therefore, a multiplier can be implemented as a single n-bit adder and a shift register.

Two's complement numbers present a particular difficulty. It would be possible, but undesirable, to recode the operands as unsigned numbers with a sign bit. Booth's algorithm tackles the problem by treating an operand as a set of sequences of all 1s and all 0s. For example, –30 is represented as 100010. This is equal to $-2^5 + 2^2 - 2^1$.

In other words, as each bit is examined in turn, from left to right, only a change from a 1 to a 0 or a 0 to a 1 is significant. Hence, in multiplying b by a, each pair of bits of a is examined, so that if $a_i = 0$ and $a_{i-1} = 1$, b shifted by i places is added to the partial product. If $a_i = 1$ and $a_{i-1} = 0$, b shifted by i places is subtracted from the partial product. Otherwise, no operation is performed. The SystemVerilog model shown in the following implements this algorithm, but note that instead of shifting the operand to the left, the partial product is shifted to the right at each clock edge. A ready flag is asserted when the multiplication is complete.

```
module booth #(parameter AL = 8, BL = 8, QL = AL+BL)
  (output logic [QL-1:0] qout, output logic ready,
   input logic [AL-1:0]ain, input logic [BL-1:0] bin,
   input logic clk, load, n_reset);

  logic [clog2(AL):0] count;
  logic [BL-1:0] alu_out;
  logic a_1;

  function int clog2(input int n);
    begin
    clog2 = 0;
    n--;
    while (n > 0)
      begin
          clog2++;
          n >>= 1;
          end
    end
  endfunction

always_ff @(posedge clk, negedge n_reset)
  if (~n_reset)
    begin
    qout <= '0;
    a_1 <= '0;
    end
  else if (load)
    begin
    qout <= ain;
    a_1 <= '0;
    end
  else if (count > 0)
    begin
    a_1 <= qout[0];
```

```
    qout <= {alu_out[BL-1],alu_out[BL-1:0],
             qout[AL-1:1]};
    end

always_ff @(posedge clk, negedge n_reset)
  if (~n_reset)
    count <= 0;
  else if (load)
    count <= AL;
  else
    count <= count - 1;

always_comb
  case ({qout[0], a_1})
    2'b01: alu_out = qout[QL-1:AL] + bin;
    2'b10: alu_out = qout[QL-1:AL] - bin;
    default: alu_out = qout[QL-1:AL];
  endcase

always_comb
  if (~load & !count)
    ready = '1;
  else
    ready = '0;

endmodule
```

5.8 Testbenches for Sequential Building Blocks

In the previous chapter, we looked at how testbenches for combinational circuits can be designed. Here and in the next chapter, we consider testbenches for sequential circuits. In this section, we consider clock generation, modeling asynchronous resets, and other deterministic signals. We also look at collecting responses. In the next chapter, we extend these ideas to synchronization with the clock.

5.8.1 Clock Generation

The most important signal in any design is the clock. In the simplest case, a clock can be generated by inverting its value at a regular interval.

The default value of any signal is "x." Simply inverting a signal at a regular interval will invert the "x" value. Thus, the following will not work; the clock would stay at "x":

```
assign #10ps clock = ~clock;
```

Therefore, the signal has to be initialized. This could be done by using an initial procedure:

```
initial clock = '0;

always #10ps clock = ~clock;
```

This explicitly uses the initial procedure as an initialization. In this case, the approach will work, but in general this is a poor coding style. The clock signal is driven from two procedures. We cannot be certain in which order procedures will be executed. Some simulators will evaluate procedures in order of declaration; other simulators will evaluate all the initial procedures first. It is far better to drive each signal from exactly one procedure. An example of this is

```
initial
  begin
  clock = '0;
  forever #10ps clock = ~clock;
  end
```

This could also be done by assigning specific values to the clock.

```
always
  begin
  #10ps clock = '0;
  #10ps clock = '1;
  end
```

All of these clock generation examples model a clock with equal high and low periods. The following example shows a clock generator in which the frequency and mark/space ratio are parameters. Notice that (a) the time precision is specified to be one-tenth of the time unit and (b) the clock frequency and mark period are specified as real numbers. Both of these conditions must be fulfilled for the example given to simulate correctly. If the frequency were specified as an integer, a mark period of 45% will cause a clock to be generated with a period of 9 ns, and mark and space times of 4 ns and 5 ns, respectively, because of rounding errors.

```
module Clockgen;
timeunit 1ns;
timeprecision 100ps;
parameter ClockFreq_MHz = 100.0; // 100 MHz
parameter Mark = 45.0; // Mark length %
// Mark time in ns
parameter ClockHigh = (Mark*10)/ClockFreq_MHz;

// Space time in ns
parameter ClockLow = ((100 - Mark)*10)/ClockFreq_MHz;
```

```
logic clock;

initial
  begin
  clock = '0;
  forever
    begin
    #ClockLow clock = '1;
    #ClockHigh clock = '0;
    end
  end

endmodule
```

5.8.2 Reset and Other Deterministic Signals

After the clock, the next most important signal is probably the reset (or set). The clock generation process repeats, but the reset signal is usually only asserted once at the beginning of a simulation, so an initial statement is used.

```
initial
  begin
       n_reset = '1;
  #1ns n_reset = '0;
  #1ns n_reset = '1;
  end
```

This is exactly the same, in form, as the signal generation process for combinational circuits as given in the previous chapter. Note that the reset is de-asserted at the start of the simulation and asserted a short time later. This is to ensure that the state of the circuit prior to the reset can be observed. In exactly the same way, other deterministic waveforms can be generated.

```
initial
  begin
        data = 4'b1111;
  #10ns data = 4'b0010;
  #20ns data = 4'b1101;
  #5ns  data = 4'b0000;
  end
```

5.8.3 Checking Responses

In the previous chapter, we saw how an error signal can be generated if a combinational model behaves in a different way from that expected.

The simplest, simulator-independent way to monitor what is happening is to write messages to the user. For example, the following procedure is executed whenever the output of the counter of Section 5.5.1 changes.

```
always @(count)
  $display("%t Counter has value %d", $time, count);
```

The *system task* **$display** writes text in a similar way to *printf* in C. The *system function* **$time** returns the current simulation time. There are two system tasks: **$display** and **$write** for generating general textual output. The difference between them is that **$display** automatically includes a new line character, while **$write** does not. %t is an example of a format specifier.

Table 5.7 lists all of the format specifiers. Note that either upper or lower case specifiers may be used (e.g., %t and %T are equivalent).

All specifiers appear in a string and (except for %m) require a parameter following in the **$display** or **$write** call. The data is right justified unless a format specifier is included. Except for real numbers, only the value 0 may be used, which suppresses leading spaces (e.g., %0o). Real numbers may be formatted as in C (e.g., %10.3f prints a number in 10 places with 3 fractional places.)

In the first string parameter, there can be a number of special characters as shown in Table 5.8.

Two other output tasks allow signals to be displayed: **$monitor** and **$strobe**. While **$display** and **$write** generate outputs at exactly the point at which they are called in the simulation cycle, **$monitor** outputs data continuously while **$strobe** only outputs data at the end of the simulation cycle. Only one **$monitor** process

Table 5.7 Format Specifiers

Specifier	Meaning
%h	Hexadecimal format
%d	Decimal format
%o	Octal format
%b	Binary format
%c	ASCII character format
%v	Net signal strength
%m	Hierarchical name of current scope
%s	String
%t	Time
%e	Real in exponential format
%f	Real in decimal format
%g	Real in exponential or decimal format

Table 5.8 Special Characters

Symbol	Meaning
\n	New line
\t	Tab
\\	\character
\"	" character
\xyz	Where xyz are octal digits—the character given by that octal code
%%	% character

can be active at a time. Every time one of the arguments to the monitor task changes, a new set of data is displayed. On the other hand, the **$strobe** task will only display stable data—the state of signals at the end of the cycle, just before moving to the next simulation time.

Summary

In this chapter, we discussed a number of common sequential building blocks. SystemVerilog models of these blocks have been written using processes. Most of these models are synthesizable using RTL synthesis tools. We also considered examples of testbenches for sequential circuits.

Further Reading

As with combinational blocks, manufacturers' data sheets are a good source of information about typical devices. In particular, it is worthwhile to look in some detail at the timing specifications for SRAM and DRAM devices. The multiplier is an example of how relatively complicated computer arithmetic can be performed. Hennessey and Patterson [9] have a good description of computer arithmetic units.

Exercises

5.1 Show how positive edge-triggered behavior can be described in SystemVerilog.

5.2 Write a behavioral SystemVerilog model of a negative edge-triggered D flip-flop with asynchronous active-low set and synchronous active-high reset.

5.3 Write a SystemVerilog model of a negative edge-triggered T-type flip-flop.

5.4 Write a SystemVerilog model of a 10-state synchronous counter that asserts an output when the count reaches 10.

5.5 Write a SystemVerilog model of an N-bit counter with a control input "Up." When the control input is 1, the counter counts up; when it is 0, the counter counts down. The counter should not, however, wrap round. When the all 1s or all 0s states are reached, the counter should stop.

5.6 Write a SystemVerilog model of an N-bit parallel to serial converter.

5.7 Write a SystemVerilog testbench for this parallel to serial converter.

5.8 What are the advantages and disadvantages of ripple counters as opposed to synchronous counters?

5.9 Design a synchronous Johnson counter that visits eight distinct states in sequence. How would this counter be modified such that any unused states lead, eventually, to the normal counting sequence?

5.10 Design an LFSR that cycles through the following states:
001,010,101,011,111,110,100
Verify your design by simulation.

5.11 Explain the function of the device shown in Figure 5.14. Your answer should include a description of all of the symbols.

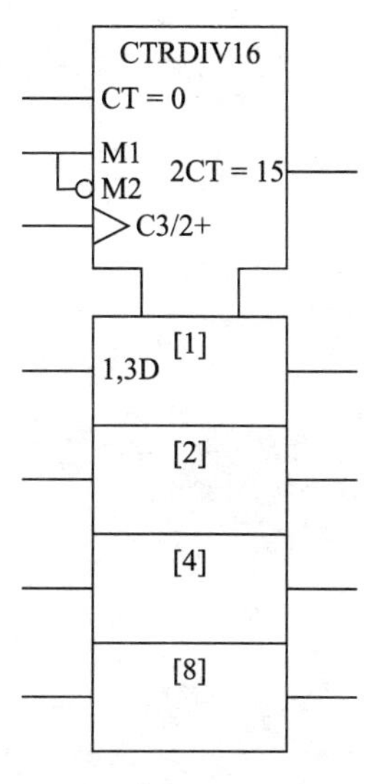

Figure 5.14 Device for Exercises 5.11 and 5.12.

5.12 Show, with a full circuit diagram, how the device of Figure 5.14 could be used to build a synchronous counter with 12 states. Show how a synchronous reset can be included.

6 Synchronous Sequential Design

Thus far we have looked at combinational and sequential building blocks. Real digital systems are usually synchronous sequential systems. In this chapter we explain how general synchronous sequential systems are designed. We then describe how such systems may be modeled in SystemVerilog.

6.1 Synchronous Sequential Systems

Almost all large digital systems have some concept of state built into them. In other words, the outputs of a system depend on past values of its inputs as well as the present values. Past input values either are stored explicitly or cause the system to enter a particular state. Such systems are known as *sequential* systems, as opposed to *combinational* systems. A general model of a sequential system is shown in Figure 6.1. The present state of the system is held in the registers—hence, the outputs of the registers give the value of the present state, and the inputs to the registers will be the next state.

The present state of the system can be updated as soon as the next state changes, in which case the system is said to be *asynchronous*, or the present state can be updated only when a clock signal changes, which is *synchronous* behavior. In this chapter, we describe the design of synchronous systems. In general, synchronous design is easier than asynchronous design, so we will leave discussion of the latter topic until Chapter 13.

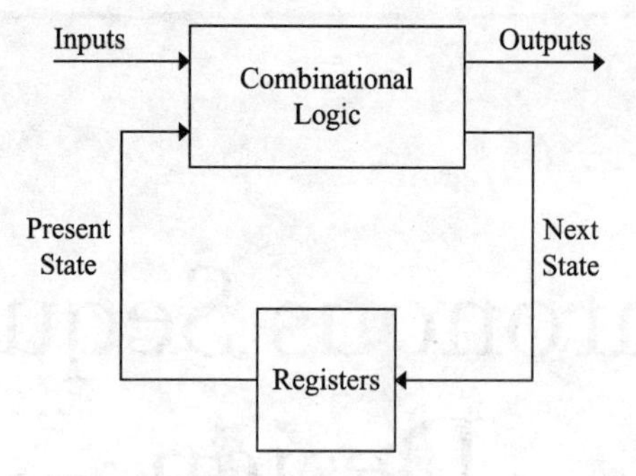

Figure 6.1 General sequential system.

In this chapter we consider the design of synchronous sequential systems. Many real systems are too complex to design in this way; thus, in Chapter 7 we show that more complex designs can be partitioned. Nevertheless, the formal design methods described in this chapter must be applied to at least part of the design of larger systems. In the next section, we introduce, by way of a simple example, a method of formally specifying such systems. We then go on to describe the problems of state assignment, state minimization and the design of the next state, and output logic. Throughout we illustrate how designs can also be modeled using SystemVerilog.

6.2 Models of Synchronous Sequential Systems

6.2.1 Moore and Mealy Machines

There are two common models of synchronous sequential systems: the *Moore* machine and the *Mealy* machine. These are illustrated in Figure 6.2. Both types of system are triggered by a single clock. The next state is determined by some (combinational) function of the inputs and the present state. The difference between the two models is that in the Moore machine, the outputs are solely a function of the present state, while in the Mealy machine, the outputs are a function of the present state and the inputs. Both the Moore and Mealy machines are commonly referred to as *state machines*. That is to say, they have an internal state that changes.

6.2.2 State Registers

As was seen in Chapter 2, combinational logic can contain hazards. The next state logic of the Moore and Mealy machines is simply a block of combinational logic with a number of inputs and a number of outputs. The existence of hazards in this next state logic could cause the system to go to an incorrect state. There are two ways

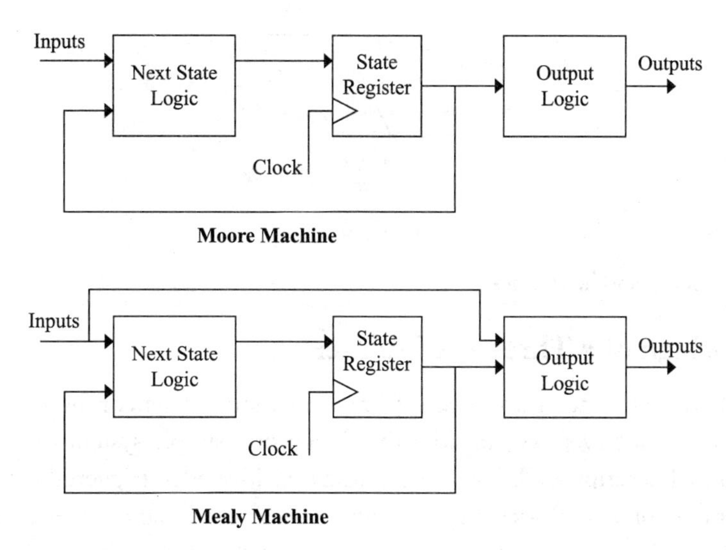

Figure 6.2 Moore and Mealy machines.

to avoid such a problem: either the next state logic should include the redundant logic needed to suppress the hazard or the state machine should be designed such that a hazard is allowed to occur but is ignored. The first solution is not ideal, as the next state logic is more complex; hence, the second approach is used. (Note that *asynchronous* systems are susceptible to hazards, and the next state logic *must* prevent any hazards from occurring, which is one reason why synchronous systems are usually preferred.)

To ensure that sequential systems are able to ignore hazards, a clock is used to synchronize data. When the clock is invalid, any hazards that occur can be ignored. A simple technique, therefore, is to logically AND a clock signal with the system signals—when the clock is at logic 0, any hazards would be ignored. The system is, however, still susceptible to hazards while the clock is high. It is common, therefore, to use registers that are only sensitive to input signals while the clock is changing. The clock edge is very short compared with the period of the clock. Therefore, the data only has to be stable for the duration of the clock change, with small tolerances before and after the clock edge. These timing tolerance parameters are known as the *setup* and *hold times* (t_{SETUP}, t_{HOLD}), respectively, as shown in Figure 6.3.

The state registers for a synchronous state machine are therefore edge-triggered D flip-flops. Other types of flip-flop may be used to design synchronous systems, but they offer few advantages and are not common in programmable logic.

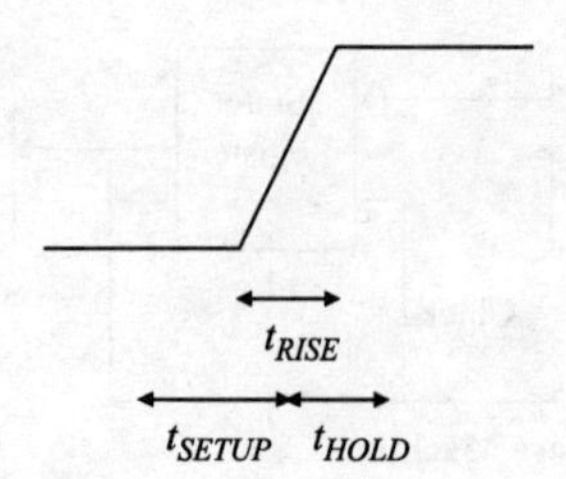

Figure 6.3 Setup and hold times.

6.2.3 Design of a Three-Bit Counter

In the next section, we introduce a formal notation for synchronous sequential systems. First, however, we consider the design of a simple system that does not need a formal description. Let us design, using positive edge-triggered D flip-flops, a counter that, on rising clock edges, counts through the binary sequence from 000 to 111, at which point it returns to 000 and repeats the sequence.

The three bits will be labeled A, B, and C. The truth table is shown in Table 6.1, in which A^+, B^+, and C^+ are the next states of A, B, and C.

A^+ etc. are the *inputs* to the state register flip-flops; A etc. are the outputs. Therefore, the counter has the structure shown in Figure 6.4. The design task is thus to derive expressions for A^+, B^+, and C^+ in terms of A, B, and C. From the truth table, K-maps can be drawn, as shown in Figure 6.5. Hence, the following expressions for the next state variables can be derived.

$$A^+ = A \cdot \bar{C} + A \cdot \bar{B} + \bar{A} \cdot B \cdot C$$
$$B^+ = B \cdot \bar{C} + \bar{B} \cdot C$$
$$C^+ = \bar{C}$$

The full circuit for the counter is shown in Figure 6.6.

Table 6.1 Truth Table of a Counter

ABC	$A^+B^+C^+$
000	001
001	010
010	011
011	100
100	101
101	110
110	111
111	000

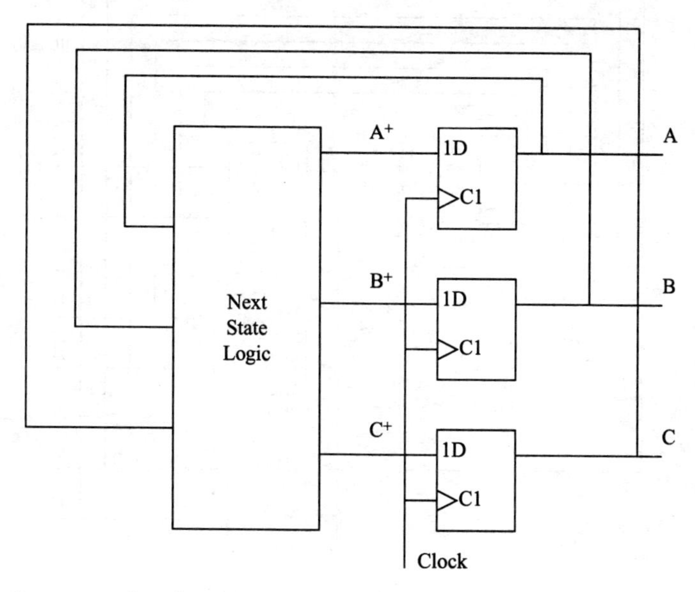

Figure 6.4 Structure of a 3-bit counter.

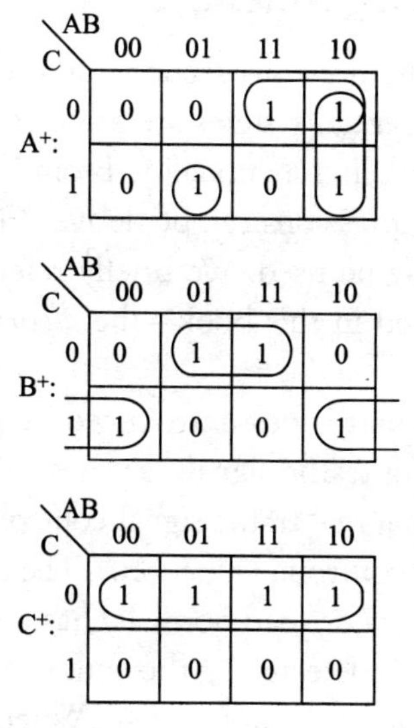

Figure 6.5 K-maps for a 3-bit counter.

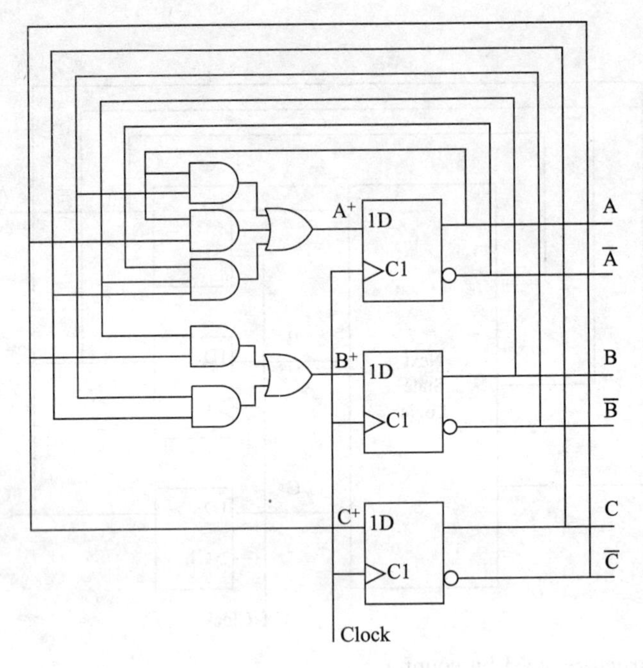

Figure 6.6 A 3-bit counter circuit.

6.3 Algorithmic State Machines

The counter designed in the previous section could easily be described in terms of state changes. Most sequential systems are more complex and require a formal notation to fully describe their functionality. From this formal notation, a state table and hence Boolean expressions can be derived. There are a number of types of formal notation that may be used. We briefly refer to one before introducing the principal technique used in this book—the *algorithmic state machine (ASM) chart*.

The form of an ASM chart is best introduced by an example. Let us design a simple controller for a set of traffic signals, as shown in Figure 6.7. This example is significantly simpler than a real traffic signal controller (and would probably be more dangerous than an uncontrolled junction!). The traffic signals have two lights each—red and green. The major road normally has a green light, while the minor road has a red light. If a car is detected on the minor road, the signals change to red for the major road and green for the minor road. When the lights change, a timer is started. Once that timer completes, a TIMED signal is asserted, which causes the lights to change back to their default state.

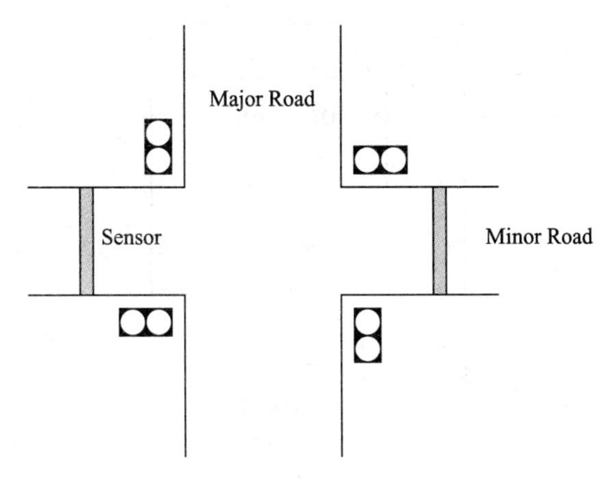

Figure 6.7 Traffic signal problem.

The functionality of this system can be described by the state machine diagram of Figure 6.8. This form of diagram is commonly used, but can be unclear. For some systems (e.g., that of Figure 12.19), such diagrams are sufficient. In this book, however, we generally use ASM charts, which are much less ambiguous. The ASM chart for the traffic signal controller is shown in Figure 6.9.

ASM charts resemble flow charts, but contain implicit timing information—the clock signal is not explicitly shown in Figure 6.9. It should be noted that ASM charts represent physical hardware. Therefore, all transitions within the ASM chart must form closed paths—hardware cannot suddenly start or stop (the only exception to this might be a reset state to which the system never returns).

The basic component of an ASM chart is the state box, shown in Figure 6.10(a). The state takes exactly one clock cycle to complete. At the top left-hand corner, the name of the state is shown. At the top right-hand corner, the state assignment

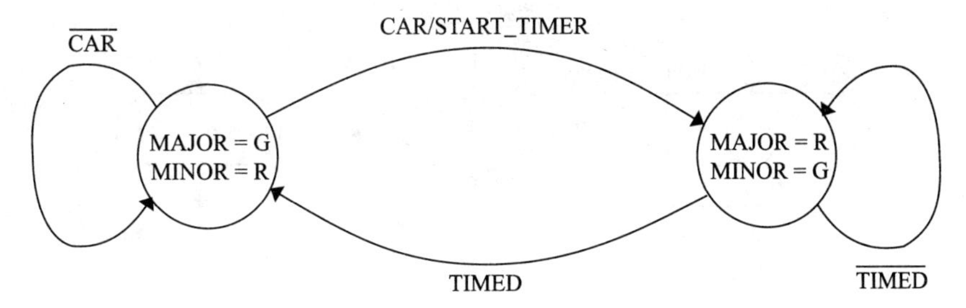

Figure 6.8 State machine of a traffic signal controller.

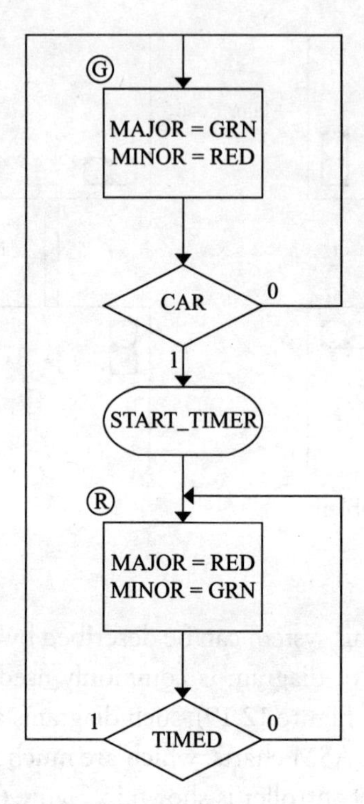

Figure 6.9 ASM chart of a traffic signal controller.

(Section 6.4.2) may be given. Within the state box, the output signals are listed. The signals take the values shown for the duration of the clock cycle and are reset to their default values for the next clock cycle. If a signal does not have a value assigned to it (e.g., Y), that signal is asserted (logic 1) during the state and is deasserted elsewhere. The notation $X \leftarrow 1$ means that the signal is assigned at the *end* of the state (i.e., during the next clock cycle) and holds its value until otherwise set elsewhere.

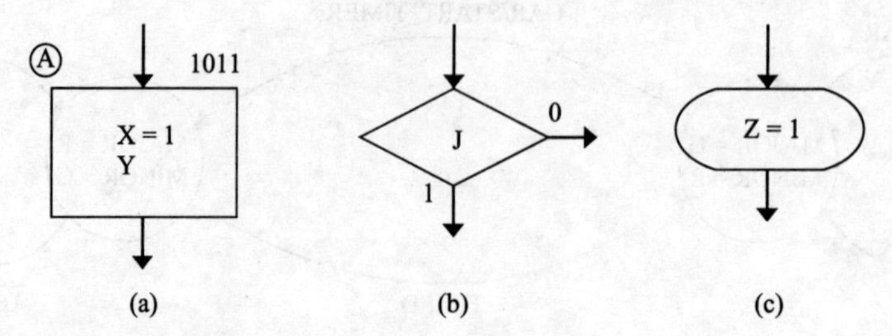

Figure 6.10 ASM chart symbols.

A decision box is shown in Figure 6.10(b). Two or more branches flow from the decision box. The decision is made from the value of one or more input signals. The decision box *must* follow and be associated with a state box. Therefore, the decision is made in the same clock cycle as the other actions of the state. Hence, the input signals must be valid at the start of the clock cycle.

A conditional output box is shown in Figure 6.10(c). A conditional output must follow a decision box. Therefore, the output signals in the conditional output box are asserted in the same clock cycle as those in the state box to which it is attached (via one or more decision boxes). The output signals can change during that state as a result of input changes. The conditional output signals are sometimes known as Mealy outputs because they are dependent on input signals, as in a Mealy machine.

It can therefore be seen that one state, or clock cycle, consists of more than just the state box. Decision boxes and conditional output boxes also form part of the state. Figure 6.9 can be redrawn, as in Figure 6.11, where all the components of a state are enclosed within dashed lines.

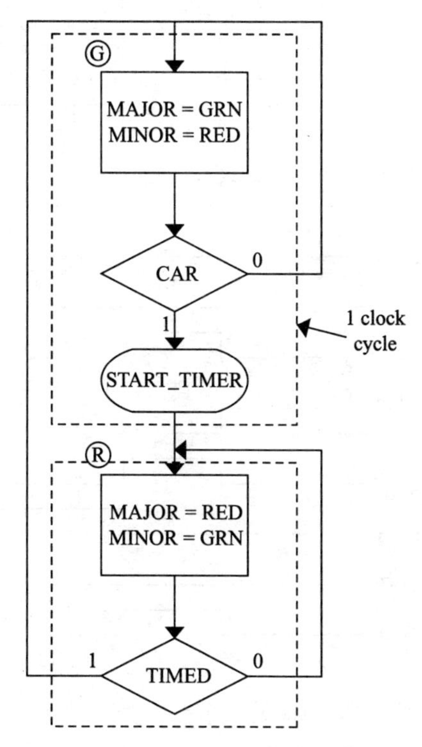

Figure 6.11 ASM chart showing clock cycles.

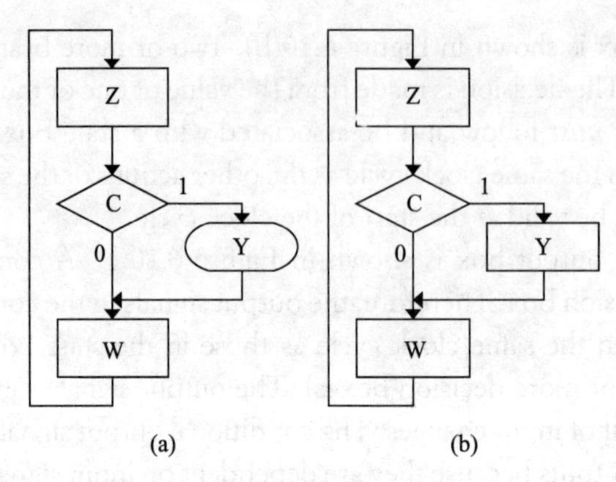

Figure 6.12 Conditional and unconditional outputs.

The difference between state boxes and conditional output boxes is illustrated in Figure 6.12. In Figure 6.12(a), there are two states. Output Y is asserted during the first state if input C is true or becomes true. In Figure 6.12(b), there are three states. The difference can be seen in the timing diagrams of Figure 6.13.

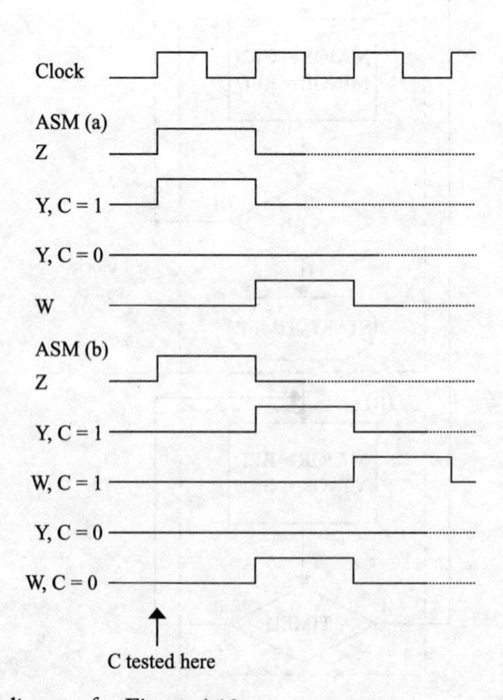

Figure 6.13 Timing diagram for Figure 6.12.

6.4 Synthesis from ASM Charts

6.4.1 Hardware Implementation

An ASM chart is a description or specification of a synchronous sequential system. It is an abstract description in the sense that it describes *what* a system does, but not *how* it is done. Any given (nontrivial) ASM chart may be implemented in hardware in more than one way. The ASM chart can, however, be used as the starting point of the hardware synthesis process. To demonstrate this, an implementation of the traffic signal controller will first be designed. We then use further examples to show how the state minimization and state assignment problems may be solved.

The ASM chart in Figure 6.9 may be equivalently expressed as a *state and output table*, as shown in Table 6.2. The outputs to control the traffic signals themselves are not shown, but otherwise the state and output table contains the same information as the ASM chart. As we will see, the state and output table is more compact than an ASM chart and is therefore easier to manipulate.

To implement this system in digital hardware, the abstract states, G and R, have to be represented by Boolean variables. Here, the problem of *state assignment* is nearly trivial. Two states can be represented by one Boolean variable. For example, when the Boolean variable, A, is 0 it can represent state G, and when it is 1, state R. It would be equally valid to use the opposite values. These values for A can be substituted into the state and output table to give the *transition and output table* shown in Table 6.3.

This transition and output table is effectively two K-maps superimposed on each other. These are explicitly shown in Figure 6.14. From these, expressions can be derived for the state variable and the output.

$$A^{+} = \bar{A} \cdot CAR + A \cdot \overline{TIMED}$$

$$START\_TIMER = \bar{A} \cdot CAR$$

Table 6.2 State and Output Table

	CAR, TIMED			
Present State	00	01	11	10
G	G,0	G,0	R,1	R,1
R	R,0	G,0	G,0	R,0
	Next State, START_TIMER			

Table 6.3 Transition and Output Table

A	CAR, TIMED 00	01	11	10
0	0,0	0,0	1,1	1,1
1	1,0	0,0	0,0	1,0
	A^+, START_TIMER			

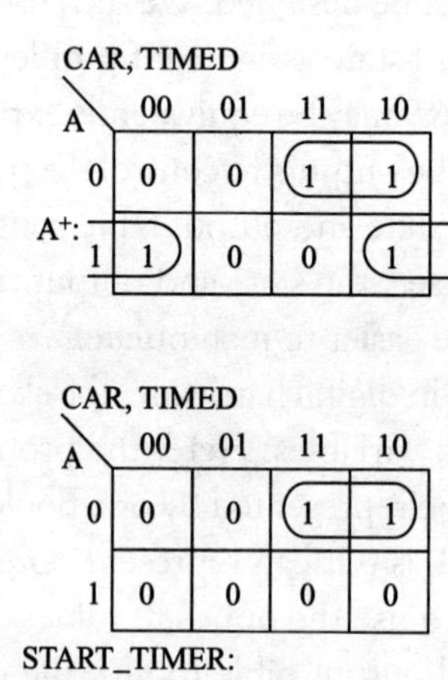

Figure 6.14 K-maps for a traffic signal controller.

For completeness, a hardware implementation is shown in Figure 6.15. The two flip-flop outputs can be used directly to control the traffic signals, so that when A is 1 (and $\bar{A}$ is 0), the signal for the major road is green, and the signal for the minor road is red. When A is 0, the signals are reversed.

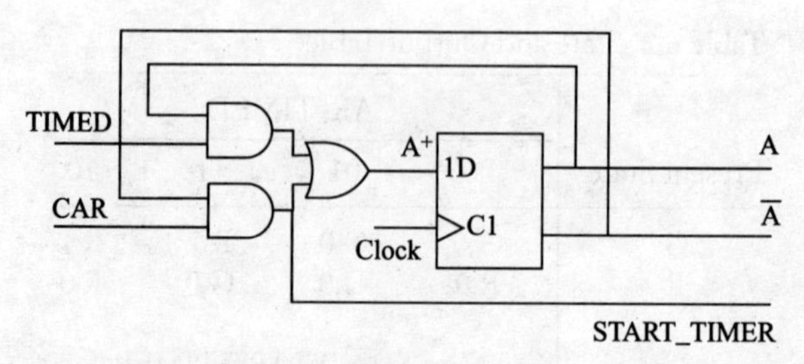

Figure 6.15 Circuit for a traffic signal controller.

6.4.2 State Assignment

In the previous example, there were two possible ways to assign the abstract states, G and R, to the Boolean state variable, A. With more states, the number of possible state assignments increases. In general, if we want to code s states using a minimal number of D flip-flops, we need m Boolean variables, where $2^{m-1} < s \leq 2^m$. The number of possible assignments is given by

$$\frac{(2^m)!}{(2^m - s)!}.$$

This means, for example, that there are 24 ways to encode three states using two Boolean variables and 6720 ways to encode five states using three Boolean variables. In addition, there are possible state assignments that use more than the minimal number of Boolean variables, which may have advantages under certain circumstances. There is no known method for determining in advance which state assignment is "best" in the sense of giving the simplest next state logic. It is obviously impractical to attempt every possible state assignment. Therefore, a number of *ad hoc* guidelines can be used to perform a state assignment. Again, let us use an example to demonstrate this.

A synchronous sequential system has two inputs, X and Y, and one output, Z. When the sum of the inputs is a multiple of 3, the output is true—it is false otherwise. The ASM chart is shown in Figure 6.16.

To encode the three states, we need (at least) two state variables and hence two flip-flops. As noted previously, there are 24 ways to encode three states; which should we use? We could arbitrarily choose any one of the possible state assignments, or we could apply one or more of the following guidelines.

- It is good practice to provide some means of initializing the state machine when power is first applied. This can be done using the asynchronous resets or sets on the system flip-flops. Therefore, the first state (state A in this example) can be coded as all 0s or all 1s.
- We can use the normal binary counting sequence for further states (e.g., B becomes 01 and C becomes 10).
- We can minimize the number of bits that change between states, for example, by using a Gray code. (This doesn't help in this example as transitions exist from each state to every other state.)
- The states might have some particular meaning. Thus, a state variable bit might be set in one state but in no others. (This can result in a non-minimal

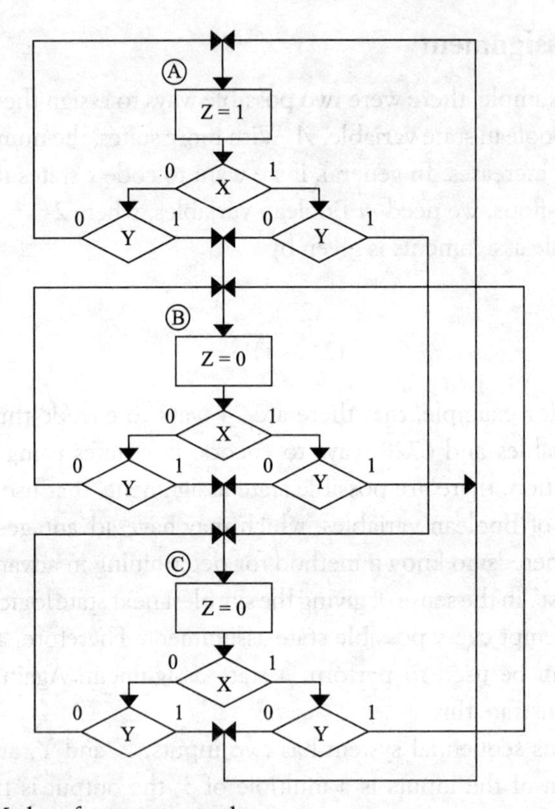

Figure 6.16 ASM chart for a sequence detector.

number of state variables but very simple output logic, which under some circumstances can be very desirable.)

- We can use one variable per state. For three states, we would have three state variables and hence three flip-flops. The states would be encoded as 001, 010, and 100. This is known as "one-hot" encoding, as only one flip-flop is asserted at a time. Although this appears to be very non-optimal, there may be advantages to the one-hot (or "one-cold") method. The next state logic may be relatively simple. In some forms of programmable logic, such as FPGAs, there is a very high ratio of flip-flops to combinational logic. A one-hot encoded system may therefore use fewer resources than a system with a minimal number of flip-flops. Furthermore, because exactly one flip-flop output is asserted at a time, it is relatively easy to detect a system malfunction in which this condition is not met. This can be very useful for safety-critical systems.

Let us therefore apply a simple state encoding to the example. The state and output table is shown in Table 6.4, and the transition and output table is shown in

Table 6.4 State and Output Table for Sequence Detector

	X, Y			
P	00	01	11	10
A	A,1	B,1	C,1	B,1
B	B,0	C,0	A,0	C,0
C	C,0	A,0	B,0	A,0
	P^+, Z			

Table 6.5 Transition and Output Table for Sequence Detector

	X,Y			
S_1S_0	00	01	11	10
00	00,1	01,1	11,1	01,1
01	01,0	11,0	00,0	11,0
11	11,0	00,0	01,0	00,0
	$S_1^+S_0^+$, Z			

Table 6.5, where state A is encoded as 00, B as 01, and C as 11. The combination 10 is not used.

The fact that we have one or more unused combinations of state variables may cause a problem. These unused combinations are states of the system. In normal operation, the system would never enter these "unused states." Therefore, in principle, we can treat the next state and output values as don't cares, as shown in Figure 6.17.

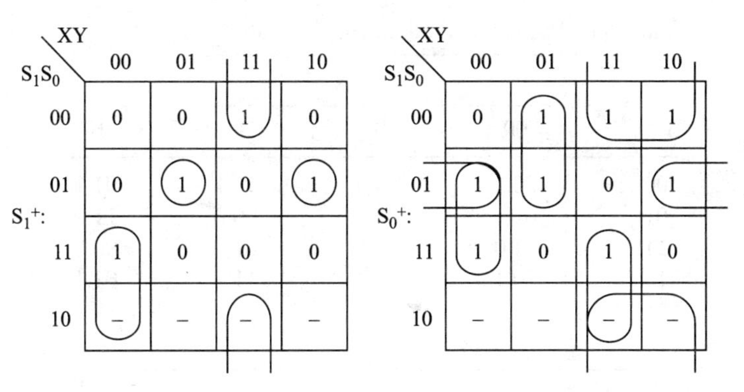

Figure 6.17 K-maps with don't cares.

This gives the next state equations:

$$S_1^+ = S_1 \cdot \bar{X} \cdot \bar{Y} + \overline{S_1} \cdot S_0 \cdot \bar{X} \cdot Y + \overline{S_0} \cdot X \cdot Y + \overline{S_1} \cdot S_0 \cdot X \cdot \bar{Y}$$
$$S_0^+ = S_0 \cdot \bar{X} \cdot \bar{Y} + \overline{S_1} \cdot \bar{X} \cdot Y + \overline{S_0} \cdot X + \overline{S_1} \cdot S_0 \cdot \bar{Y} + S_1 \cdot X \cdot Y$$

The output expression can be read directly from the transition and output table:

$$Z = \overline{S_0}$$

By default, therefore, the transitions from the unused state have now been defined, as shown in Table 6.6. Although this unused state should never be entered, it is possible that a "non-logical" event, such as a glitch on the power supply, might cause the system to enter this unused state. It can be seen from Table 6.6 that if, for example, the inputs were both 0, the system would stay in the unused state. In the worst case, once having entered an unused state, the system might be stuck in one or more unused states. The unused states could therefore form a "parasitic" state machine (or perhaps a "parallel universe"!), causing the system to completely malfunction. We could, reasonably, decide that the chances of entering an unused state are so low as to be not worth worrying about. Hence, we treat the transition table entries for the unused states as don't cares, as shown, which minimizes the next state logic. On the other hand, the system might be used in a safety-critical application. In this case, it might be important that all transitions from unused states are fully defined, so that we can be certain to return to normal operation as soon as possible. In this case, the transitions from the unused state would not be left as don't cares in the K-maps, but would be explicitly set to lead to, say, the all 0s state.

Table 6.6 Transition Table Implied by Don't Cares

	X,Y			
S_1S_0	00	01	11	10
00	00,1	01,1	11,1	01,1
01	01,0	11,0	00,0	11,0
11	11,0	00,0	01,0	00,0
10	10,1	00,1	11,1	01,1
	$S_1^+S_0^+$, Z			

Hence, the X entries in the K-maps of Figure 6.17 become 0s, and the next state equations would be:

$$S_1^+ = S_1 \cdot S_0 \cdot \bar{X} \cdot \bar{Y} + \overline{S_1} \cdot S_0 \cdot \bar{X} \cdot Y + \overline{S_1} \cdot \overline{S_0} \cdot X \cdot Y + \overline{S_1} \cdot S_0 \cdot X \cdot \bar{Y}$$
$$S_0^+ = S_0 \cdot \bar{X} \cdot \bar{Y} + \overline{S_1} \cdot \bar{X} \cdot Y + \overline{S_1} \cdot \overline{S_0} \cdot X + \overline{S_1} \cdot S_0 \cdot \bar{Y} + S_1 \cdot S_0 \cdot X \cdot Y$$

These equations are more complex than the previous set that includes the don't cares; hence, the next state logic would be more complex.

Therefore, we have a choice: We can either assume that it is impossible to enter an unused state and minimize the next state equations by assuming the existence of don't cares or we can try to reduce the risk of becoming stuck in an unused state by explicitly defining the transitions from the unused states and hence have more complex next state logic.

6.4.3 State Minimization

In the previous section we noted that to encode s states we need m flip-flops, where $2^{m-1} < s \leq 2^m$. If we can reduce the number of states in the system, we *might* reduce the number of flip-flops, hence making the system simpler. Such savings may not always be possible. For instance, the encoding of 15 states requires four flip-flops. If we reduced the number of states to nine, we would still need four flip-flops. So there would be no obvious saving and we would have increased the number of unused states, with the potential problems discussed in the previous section. As will be seen, state minimization is a computationally difficult task, and in many cases, it would be legitimate to decide that there would be no significant benefits and hence the task would not be worth performing.

State minimization is based on the observation that if two states have the same outputs and the same next states, given a particular sequence of inputs, it is not possible to distinguish between the two states. Hence, the two states are considered to be *equivalent* and hence they may be merged, reducing the total number of states.

For example, let us design the controller for a ticket vending machine. A 7-day subway ticket costs $40. The machine accepts $20 and $10 bills (all other bills are rejected by the mechanics of the system). Once $40 has been inserted, the ticket is dispensed. If more than $40 is inserted, all bills are returned. The machine has two lights: one to show that it is ready for the next transaction, and one to show that further bills need to be inserted. The ASM chart has been split into two parts (Figures 6.18 and 6.19)—the connections between the two parts are shown by circles with lower case letters.

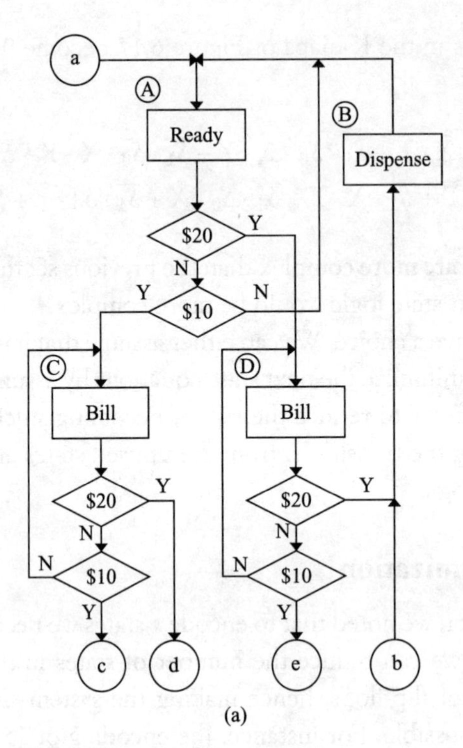

Figure 6.18 ASM chart of a vending machine (Part 1).

There are nine states in this state machine. Four flip-flops would therefore be required to implement it. If we could merge at least two states, we would save ourselves a flip-flop. From Figures 6.18 and 6.19, notice that states F, G, and H all have transitions to state I if a \$20 bill is inserted and to state B if a \$10 bill is inserted. Otherwise, all three states have transitions back to themselves. Intuitively, these three states would appear to be equivalent. Another way of looking at this is to say that states F, G, and H all represent the condition where another \$10 is expected to complete the sale of a ticket. From the point of view of the purchaser, these states are indistinguishable.

Instead of attempting to manipulate the ASM chart, it is probably clearer to rewrite it as a state and output table (Figure 6.20). The "Other" column shows the next state if no valid bill is inserted. Because there are no conditional outputs, it is possible to separate the outputs from the next state values.

The condition for two states to be considered equivalent is that their next states and outputs should be the same. States A, B, and I have unique outputs and therefore cannot be equivalent to any other states. States C to H inclusive have the

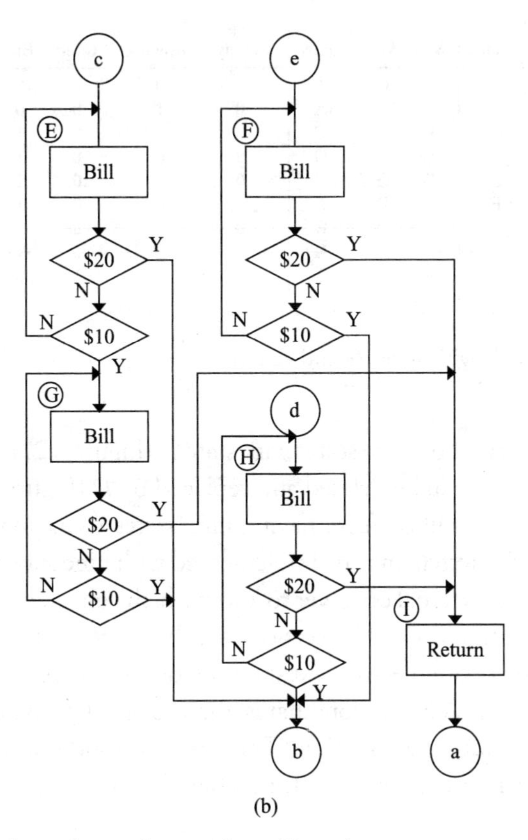

Figure 6.19 ASM chart of a vending machine (Part 2).

same outputs. States F, G, and H have the same next states, other than their default next states, which are the states themselves. In other words, states F, G, and H are equivalent if states F, G, and H are equivalent—which is a tautology! Therefore, we can merge these three states. In other words, we will delete states G and H, say, and

State	$20	$10	Other	Ready	Dispense	Return	Bill
A	D	C	A	1	0	0	0
B	A	A	A	0	1	0	0
C	H	E	C	0	0	0	1
D	B	F	D	0	0	0	1
E	B	G	E	0	0	0	1
F	I	B	F	0	0	0	1
G	I	B	G	0	0	0	1
H	I	B	H	0	0	0	1
I	A	A	A	0	0	1	0
	Next State			Outputs			

Figure 6.20 State and output table for a vending machine.

State	$20	$10	Other	Ready	Dispense	Return	Bill
A	D	C	A	1	0	0	0
B	A	A	A	0	1	0	0
C	~~H~~ F	E	C	0	0	0	1
D	B	F	D	0	0	0	1
E	B	~~G~~ F	E	0	0	0	1
F	I	B	F	0	0	0	1
~~G~~	~~I~~	~~B~~	~~G~~	~~0~~	~~0~~	~~0~~	~~1~~
~~H~~	~~I~~	~~B~~	~~H~~	~~0~~	~~0~~	~~0~~	~~1~~
I	A	A	A	0	0	1	0
	Next State				Outputs		

Figure 6.21 State table with states *G* and *H* removed.

replace all instances of those two states with state *F* (Figure 6.21). Now states *D* and *E* are equivalent, so *E* can be deleted and replaced by *D* (Figure 6.22). The system has therefore been simplified from having nine states to having six. It should be remembered that the system may be implemented with nine states or with six, but it is not possible for an external observer to know which version has been built simply by observing the outputs. The two versions are therefore functionally identical.

To conclude this example, the next state and output expressions will be written, assuming a one-hot implementation; that is, there is one flip-flop per state, of which exactly one has a 1 output at any time. The next state and output expressions can be read directly from the state and output table of Figure 6.22.

$$A^{+} = B + I + \overline{\$20} \cdot \overline{\$10} \cdot A$$

$$B^{+} = D \cdot \$20 + F \cdot \$10$$

$$C^{+} = A \cdot \$10 + \overline{\$20} \cdot \overline{\$10} \cdot C$$

$$D^{+} = A \cdot \$20 + C \cdot \$10 + \overline{\$20} \cdot \overline{\$10} \cdot D$$

$$F^{+} = C \cdot \$20 + D \cdot \$10 + \overline{\$20} \cdot \overline{\$10} \cdot F$$

State	$20	$10	Other	Ready	Dispense	Return	Bill
A	D	C	A	1	0	0	0
B	A	A	A	0	1	0	0
C	~~H~~ F	~~E~~ D	C	0	0	0	1
D	B	F	D	0	0	0	1
~~E~~	~~B~~	~~G F~~	~~E~~	~~0~~	~~0~~	~~0~~	~~1~~
F	I	B	F	0	0	0	1
~~G~~	~~I~~	~~B~~	~~G~~	~~0~~	~~0~~	~~0~~	~~1~~
~~H~~	~~I~~	~~B~~	~~H~~	~~0~~	~~0~~	~~0~~	~~1~~
I	A	A	A	0	0	1	0
	Next State				Outputs		

Figure 6.22 State table with states *E*, *G*, and *H* removed.

$$I^{+} = F \cdot \$20$$
$$Ready = A$$
$$Dispense = B$$
$$Return = I$$
$$Bill = C + D + F$$

6.5 State Machines in SystemVerilog

6.5.1 A First Example

SystemVerilog is a very rich language in terms of constructs. Therefore, there is often more than one way to describe something. Here, we will look at two styles for modeling state machines. Both are synthesizable, and both simulate. To some extent, the choice of style is a matter of taste. It should, however, be noted that some synthesis tools produce better results with one style rather than the other.

The state of the system must be held in an internal register. In SystemVerilog, the state can be represented by an *enumerated type*. The possible values of this type are the state names and the name of the variable is given after the list of values, for example,

```
enum {s0, s1, ...} state;
```

In the following listing, there are two procedural blocks. Each has a label. The first procedural block (SEQ) models the state machine itself. The procedural block waits until the clock input changes to 1, or the reset changes to 0. The asynchronous reset is tested first, and if it is asserted, a default value is assigned to the state. Otherwise, a **case** statement is used to branch according to the current value of the state. Each branch of the **case** statement is therefore equivalent to one of the states of Figure 6.9, together with its decision box. Within the first statement branch, the `car` input is tested to set the state. If the input is false, the state remains as it was (i.e., `G`). This is fine, as the block is declared to be **always_ff** and therefore we would expect the state to be mapped onto one or more edge-triggered flip-flops. The other state is structured in a similar way.

In the second procedural block (OP), the outputs are set. This is an **always_comb** block. Note that start_timer and the other outputs are given default values at the beginning. This is good practice, as it ensures that latches will not be accidentally created. Again, a **case** statement is used. The structure mirrors the

ASM chart. Unconditional outputs are assigned in each state; conditional assignments follow an **if** statement.

```
module traffic_1 (output logic start_timer,
                  major_green, minor_green,
                  input logic clock, n_reset, timed,
                  car);
  enum {G, R} state;

always_ff @(posedge clock, negedge n_reset)
  begin: SEQ
  if (~n_reset)
    state <= G;
  else
    case (state)
      G: if (car)
           state <= R;
      R: if (timed)
           state <= G;
    endcase
  end

always_comb
  begin: OP
  start_timer = '0;
  minor_green = '0;
  major_green = '0;
  case (state)
    G: begin
       major_green = '1;
       if (car)
         start_timer = '1;
       end
    R: minor_green = '1;
  endcase
  end

endmodule
```

Another common modeling style for state machines also uses two processes. One process is used to model the state registers, while the second process models the next state and output logic. The two processes therefore correspond to the two boxes in Figure 6.1. From Figure 6.1, it can be seen that the communication between the two processes is achieved using the present and next values of the state registers. Therefore, if two SystemVerilog blocks are used, communication between them must be performed using present and next state variables.

The combinatorial block (labeled COM) combines the **case** statement parts of the previous version. The **case** statement now selects on `present_state` and `next_state` is updated. Note also that `next_state` is updated (to its existing value) even when a change of state does not occur. Failure to do this would result in latches being created.

```
module traffic_2 (output logic start_timer,
                  major_green, minor_green,
                  input logic clock, n_reset, timed,
                  car);
  enum {G, R} present_state, next_state;

always_ff @(posedge clock, negedge n_reset)
  begin: SEQ
  if (~n_reset)
    present_state <= G;
  else
    present_state <= next_state;
  end

always_comb
  begin: COM
  start_timer = '0;
  minor_green = '0;
  major_green = '0;
  next_state = present_state;
  case (present_state)
    G: begin
       major_green = '1;
       if (car)
         begin
         start_timer = '1;
         next_state = R;
         end
       end
    R: begin
       minor_green = 1'b1;
       if (timed)
         next_state = G;
       end
  endcase
  end

endmodule
```

It is also possible to derive a three-process model: state register; next state evaluation, and output assignment. There is no obvious advantage to using this model. A one-process model is usually wrong—all outputs would be registered.

Some general comments apply to all styles of state machine. The inputs and outputs are of type logic. Nonblocking assignments are always used in the **always_ff** block, while blocking assignments are always used in the **always_comb** block. Never mix the two types of assignment in one block. Also note that a variable is only ever written to by one block. To avoid accidental latches, all the outputs in an **always_comb** block are initialized at the start of the block.

6.5.2 A Sequential Parity Detector

Consider the following system. Data arrives at a single input, with one new bit per clock cycle. The data are grouped into packets of 4 bits, where the fourth bit is a parity bit. (This problem could easily be scaled to have more realistically sized packets.) The system uses even parity. In other words, if there is an odd number of 1s in the first 3 bits, the fourth bit is a 1. If an incorrect parity bit is detected, an error signal is asserted during the fourth clock cycle.

The parity detector can be implemented as a state machine. We will leave the design as an exercise and simply show a SystemVerilog implementation. In this example, an asynchronous reset is included to set the initial state to s0. Notice that the error signal is only set under limited conditions, making the combinational logic block very simple.

```
module seqparity (output logic error,
                  input logic clock, n_reset, a);

  enum {s0, s1, s2, s3, s4, s5, s6} state;

always_ff @(posedge clock, negedge n_reset)
  begin: SEQ
  if (~n_reset)
    state <= s0;
  else
    case (state)
      s0: if (~a)
             state <= s1;
          else
             state <= s2;
      s1: if (~a)
             state <= s3;
          else
             state <= s4;
```

```
      s2: if (~a)
            state <= s4;
          else
            state <= s3;
      s3: if (~a)
            state <= s5;
          else
            state <= s6;
      s4: if (~a)
            state <= s6;
          else
            state <= s5;
      s5: state <= s0;
      s6: state <= s0;
    endcase
  end

always_comb
  begin: COM
  if ((state == s5 && a) || (state == s6 && ~a))
    error = '1;
  else
    error = '0;
  end

endmodule
```

6.5.3 Vending Machine

The following piece of SystemVerilog is a model of the (minimized) vending machine of Section 6.4.3. Two blocks are used. Note that here an asynchronous reset has been provided to initialize the system when it is first turned on.

```
module vending(output logic ready,dispense,ret,bill
               input logic clock,n_reset,twenty,ten);

  enum {A, B, C, D, F, I} present_state, next_state;

always @(posedge clock, negedge n_reset)
  begin: SEQ
  if (~n_reset)
    present_state <= A;
  else
    present_state <= next_state;
  end

always_comb
```

```
begin: COM
ready = '0;
dispense = '0;
ret = '0;
bill = '0;
case (present_state)
  A : begin
      ready = '1;
      if (twenty)
        next_state = D;
      else if (ten)
        next_state = C;
      else
        next_state = A;
      end
  B : begin
      dispense = '1;
      next_state = A;
      end
  C : begin
      bill = '1;
      if (twenty)
        next_state = F;
      else if (ten)
        next_state = D;
      else
        next_state = C;
      end
  D : begin
      bill = '1;
      if (twenty)
        next_state = B;
      else if (ten)
        next_state = F;
      else
        next_state = D;
      end
  F : begin
      bill = '1;
      if (twenty)
        next_state = I;
      else if (ten)
        next_state = B;
      else
        next_state = F;
      end
```

```
    I : begin
        ret = '1;
        next_state = A;
        end
  endcase
  end
endmodule
```

6.5.4 Storing Data

One (of the many) problems with the traffic light controller of Section 6.5.1 is that the minor road lights will switch to green as soon as a car is detected. This will happen even if the lights have just changed. It would be preferable if the timer were used to keep the major road lights green for a period of time. If we did this simply by asserting the start_timer signal in both states and waiting for the timed signal to appear, as follows, an arriving car could easily be missed.

```
always_comb
  begin: COM
  start_timer = '0;
  minor_green = '0;
  major_green = '0;
  next_state = present_state;
  case (present_state)
    G: begin
       major_green = '1;
       if (car && timed)
         begin
         start_timer = '1;
         next_state = R;
         end
       end
    R: begin
       minor_green = 1'b1;
       if (timed)
         start_timer = '1;
         next_state = G;
       end
  endcase
  end
```

Therefore, the fact that a car has arrived needs to be remembered in some way. This could be done by adding further states to the state machine. Alternatively, the car arrival could be stored. It is not possible to say that one approach is better than the other. We will look at the idea of using a state machine to control other hardware in Chapter 7. Meanwhile, let us consider how a simple piece of data can be stored.

In a purely simulation model, it is possible to store the state of a variable or signal in a combinational process. This is done by assigning a value in one branch of the process. As we will see in Chapter 10, when synthesized, this would inevitably lead to asynchronous latches and hence timing problems. Instead, any data that is to be stored must be explicitly saved in a register, modeled as a clocked process. Storing data in this way is exactly the same as storing a state. Therefore, separate signals are needed for the present value of the car register and for the next value. We will use more meaningful names for these signals:

```
logic car_arrived, car_waiting;
```

The car_waiting signal is updated at the same time as the present_state signal.

```
always_ff @(posedge clock, negedge n_reset)
  if (~n_reset)
    begin
    present_state <= G;
    car_waiting <= '0;
    end
  else
    begin
    present_state <= next_state;
    car_waiting <= car_arrived;
    end
```

The car_arrived signal is set or reset in the following process:

```
always_comb
  begin: car_update
  if (present_state == G && car_waiting && timed)
    car_arrived = '0;
  else if (car)
    car_arrived = '1;
  else
    car_arrived = car_waiting;
  end
```

Finally, both references to car in block com at the start of this section need to be replaced by references to car_waiting. Notice that each signal is assigned in only one block. It often helps to sketch a diagram of the system with each process represented by a box and showing all the inputs and outputs of each block. If a signal appears to be an output from two boxes, or if a signal is not an input to a block, something is not right!

6.6 Testbenches for State Machines

In the previous chapter, we considered testbenches for sequential logic. The function of such testbenches was to generate clock and reset signals and to monitor outputs. In this section, we look at how inputs to a state machine can be synchronized with the clock and how we can use **assert** statements to monitor outputs.

In exactly the same way that an RTL model can be made sensitive to the clock or to some other signal, parts of the testbench can also be made sensitive to the clock. Here we use the event control construct (@) in a context other than an always block. In the **forever** loop, the procedure waits for the rising edge of the clock, and then waits for a further 5 ns before count is incremented. This ensures that the increment does not coincide with the clock edge (assuming that the clock period is greater than 5 ns).

```
integer count;

initial
  begin
  count = 0;
  forever
    begin
    @(posedge clk);
    #5ns count++;
    end
  end
```

This example has two forms of timing control: a *delay* control (#5ns) and an *event* control (`@(posedge clk)`). Either or both forms can precede a statement; thus, we could have written:

```
@(posedge clk) #5ns count++;
```

It is also possible to make a statement sensitive to any edge by writing, for example, `@clk`. A third form of event control is the level sensitive wait statement:

```
wait (!enable) #10ns count++;
```

If enable is at 1, the flow stops until enable becomes 0. If enable is already 0, there is no delay.[1]

It is also possible to generate named events and to control the flow. For example, one unit might have the named event `trigger` (defined as shown):

```
-> trigger;
```

In another process, there is an event control sensitive to that named event:

```
@trigger count++;
```

Summary

State machines can be formally described using ASM charts. The design of a synchronous state machine from an ASM chart has a number of distinct steps: state minimization, state assignment, derivation of next state, and output logic. A SystemVerilog model of a state machine can be written that is equivalent to an ASM chart. This SystemVerilog model may be automatically synthesized to hardware using an RTL synthesis tool.

Further Reading

State machine design is a core topic in digital design and therefore covered in many textbooks. Not all books use ASM notation; many use the style of Figure 6.8. The problem of state minimization is covered in detail in books such as Hill and Peterson [10].

Exercises

6.1 Explain the difference between a Mealy machine and a Moore machine.

6.2 Describe the symbols used in an ASM diagram.

---

1. If you are familiar with VHDL, be careful! The behavior of a SystemVerilog wait statement is different than a VHDL wait until statement.

6.3 The following code shows part of a SystemVerilog description of a synchronous state machine. Complete the description by writing down the synchronization process. How would an asynchronous reset be included?

```
module state_machine (output logic z,
                      input logic x, clock);

enum {S0, S1, S2, S3} state, next_state;

// synchronization statements go here!

always_comb
  begin: com
  case (state)
    S0: begin
        z = '0;
        if (X)
          next_state = S2;
        else
          next_state = S0;
        end
    S1: begin
        z = '1;
        if (x)
          next_state = S2;
        else
          next_state = S0;
        end
    S2: begin
        z = '0;
        if (x)
          next_state = S3;
        else
          next_state = S2;
        end
    S3: begin
        z = '0;
        if (x)
          next_state = S1;
        else
          next_state = S3;
        end
    endcase
  end

endmodule
```

6.4 Draw the ASM chart that describes the state machine shown in Exercise 6.3.

6.5 Draw an ASM chart to describe a state machine that detects a sequence of three logic 1s occurring at the input and that asserts a logic 1 at the output during the last state of the sequence. For example, the sequence 001011101111 would produce an output 000000100011. Write a SystemVerilog description of the state machine.

6.6 Write a testbench to stimulate the state machine of Exercise 6.5 and verify the SystemVerilog model by simulation.

6.7 Produce next state and output logic for the state machine of Exercise 6.5 and write a SystemVerilog description of the hardware using simple gates and positive edge-triggered D flip-flops. Verify this hardware by simulation.

6.8 A state machine has two inputs, A, B, and one output, Z. If the sequence of input pairs: A = 1 B = 1, A = 1 B = 0, A = 0 B = 0 is detected, Z becomes 1 during the final cycle of the sequence; otherwise, the output remains at 0. Write a SystemVerilog model of a state machine to implement this system, using two procedural blocks, such that one block models the state machine and the other models the output logic.

6.9 Rewrite the model of Exercise 6.8 to use two procedural blocks: one for the registers and one for the next state logic and output logic.

6.10 Using an ASM chart, design a traffic signal controller for a crossroads. The signals change only when a car is detected in the direction with a red signal. The signals change in the sequence: red, yellow, green, red. Note that while the signals in one direction are green, or yellow, the signals in the other direction are red (i.e., you need more than three states). Design an implementation that uses a minimal number of D flip-flops.

6.11 A counter is required to count people entering and leaving a room. The room has a separate entrance and exit. Sensors detect people entering and leaving. Up to seven people are allowed in the room at one time. Draw an ASM chart of a synchronous counter that counts the people in the room and that indicates when the room is empty and full. One person may enter and one person may leave during each clock cycle. The empty and full indicators should be asserted immediately when the condition is true, that is, before the next clock edge. Write a SystemVerilog model of the system.

6.12 Construct a state and output table for the state machine represented by Figure 6.23. Show that the number of states can be reduced. Derive the next

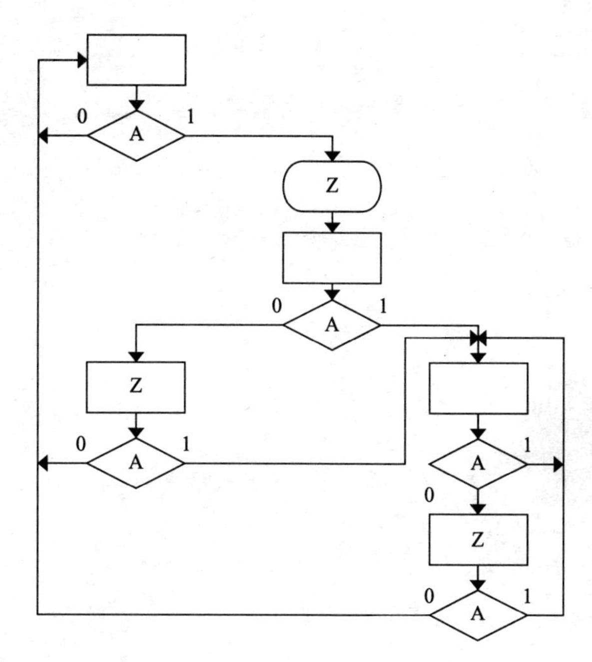

Figure 6.23 ASM chart for Exercise 6.12.

state and output logic to implement the reduced state machine using (a) a minimal number of D flip-flops and (b) the one hot D flip-flop method. What are the relative advantages of each method? How has the reduction in the number of states helped in each case?

7 Complex Sequential Systems

In the previous three chapters we looked at combinational and sequential building blocks and the design of state machines. The purpose of this chapter is to see how these various parts can be combined to build complex digital systems.

7.1 Linked State Machines

In principle, any synchronous sequential system could be described by an ASM chart. In practice, this does not make sense. The states of a system, such as a microprocessor, include all the possible values of all the data that might be stored in the system. Therefore, it is usual to partition a design in some way. In this chapter, we show first how an ASM chart, and hence the SystemVerilog model of the state machine, can be partitioned, and second how a conceptual split may be made between the *datapath* of a system, that is, the components that store and manipulate data, and the state machine that controls the functioning of those datapath components.

A large body of theory covers the optimal partitioning of state machines. In practice, it is usually sufficient to identify components that can easily be separated from the main design and implemented independently. For example, let us consider again the traffic signal controller.

If a car approaches the traffic signals on the minor road, a sensor is activated that causes the major road to have a red light and the minor road to have a green

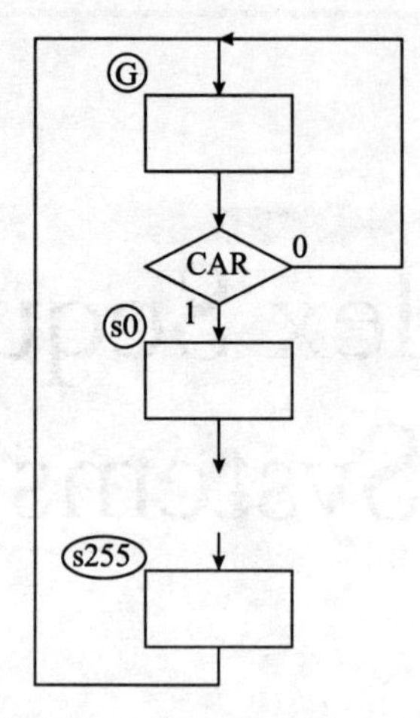

Figure 7.1 ASM chart of a traffic signal controller, including the timer.

light for a fixed interval. Once that interval has passed, the major road has a green light again and the minor road has a red light. In Chapter 6, we simply assumed that a signal would be generated after the given interval had elapsed. Let us now assume that the clock frequency is such that the timed interval is completed in 256 clock cycles. We can draw an ASM chart for the entire system as shown in Figure 7.1 (states 1 to 254 and the outputs are not shown, for clarity). Although this is a simple example, the functionality of the system is somewhat lost in the profusion of states that implement a simple counting function. It would be clearer to separate the traffic light controller function from the timer.

One way of doing this is shown in Figure 7.2, in which there are two ASM charts. The ASM chart on the left is the traffic light controller, in which a signal, START, is asserted as a conditional output when a car is detected. This signal acts as an input to the second state machine, allowing that state machine to move from the IDLE state into the counting sequence. When the second state machine completes the counting sequence, the signal TIMED is asserted, which acts as an input to the first state machine, allowing the latter to move from state *R* to state *G*. The second state machine moves back into the IDLE state.

A state machine of the form of the second state machine of Figure 7.2 can be thought of as a "hardware subroutine." In other words, any state machine may be partitioned in this way. Unlike a software subroutine, however, a piece of hardware must exist and must be doing something, even when it is not being used. Hence, the IDLE state must be included to account for the time when the "subroutine" is not doing a useful task.

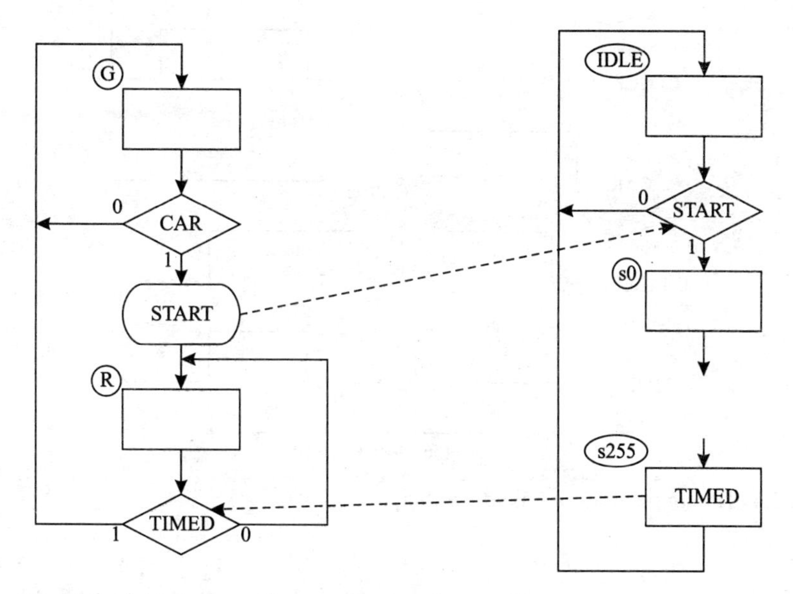

Figure 7.2 Linked ASM charts for a traffic signal controller.

An alternative way to implement a subsidiary state machine is shown in Figure 7.3. This version does not correspond to the hardware subroutine model, but represents a conventional counter. The use of standard components is discussed further in the next section.

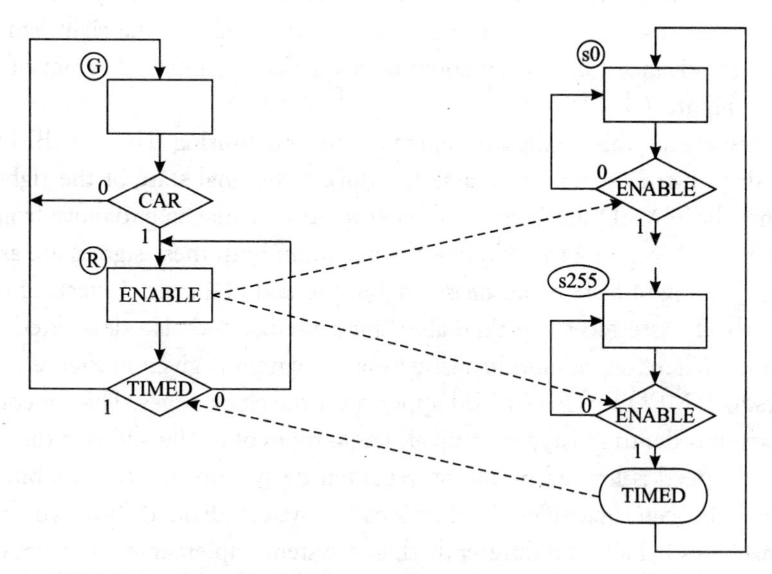

Figure 7.3 ASM chart of a traffic signal controller with counter.

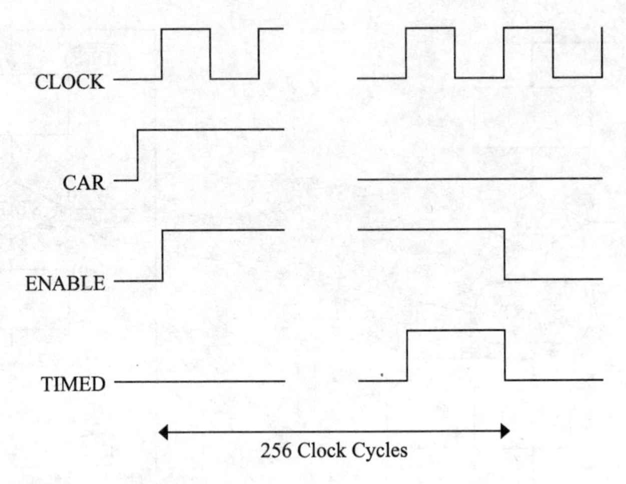

Figure 7.4 Timing diagram of linked ASM charts.

From the ASM chart of Figure 7.1, it is quite clear that the system takes 256 clock cycles to return to state *G* after a car has been detected. The sequence of operations may be harder to follow in Figure 7.3. In state *s*255, TIMED is asserted as a conditional output. This causes the left-hand state machine to move from state *R* to state *G*. In state *R*, ENABLE is asserted, which allows the right-hand state machine to advance through its counting sequence. A timing diagram of this is shown in Figure 7.4.

At first glance, this timing diagram may appear confusing. The ENABLE signal causes the TIMED signal to be asserted during the final state of the right-hand diagram. The TIMED signal causes the left-hand state machine to move from state *R* to state *G*. According to ASM chart convention, both these signals are asserted at the beginning of a state and deasserted at the end of a state. In fact, of course, the signals are asserted some time after a clock edge and also deasserted after a clock edge. Therefore, a more realistic timing diagram is given in Figure 7.5. The changes to TIMED and ENABLE happen after the clock edges. This, of course, is necessary in order to satisfy the setup and hold times of the flip-flops in the system. The clock speed is limited by the propagation delays through the combinational logic of both state machines. In that sense, a system divided into two or more state machines behaves no differently than a system implemented as a single state machine.

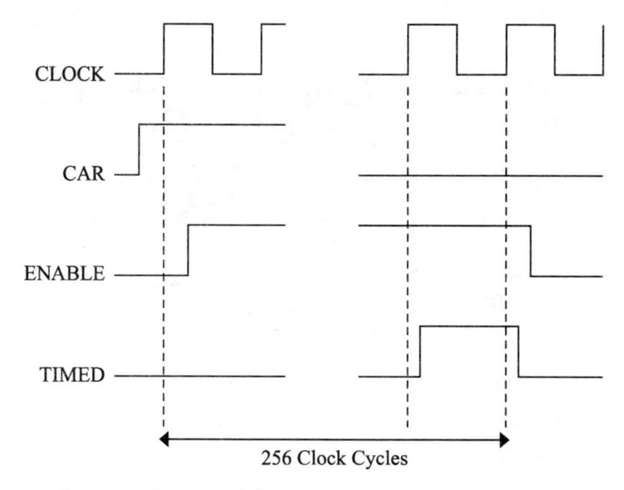

Figure 7.5 Timing diagram showing delays.

7.2 Datapath/Controller Partitioning

Although any synchronous sequential system can be designed in terms of one or more state machines, in practice this is likely to result in the "reinvention of the wheel" on many occasions. For example, the right-hand state machine of Figure 7.3 is simply an 8-bit counter. Given this, it is obviously more effective to reuse an existing counter, either as a piece of hardware or as a SystemVerilog model. It is therefore convenient to think of a sequential system in terms of the *datapath*, that is, those components that have been previously designed (or that can be easily adapted) and that can be reused, and the *controller*, which is a design-specific state machine. A model of a system partitioned in this way is shown in Figure 7.6.

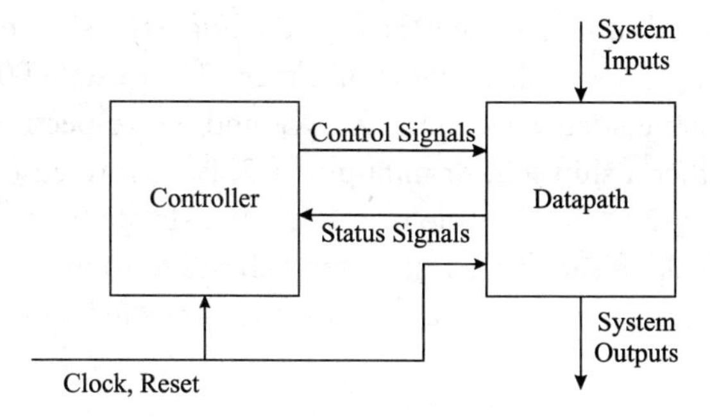

Figure 7.6 Controller/datapath partitioning.

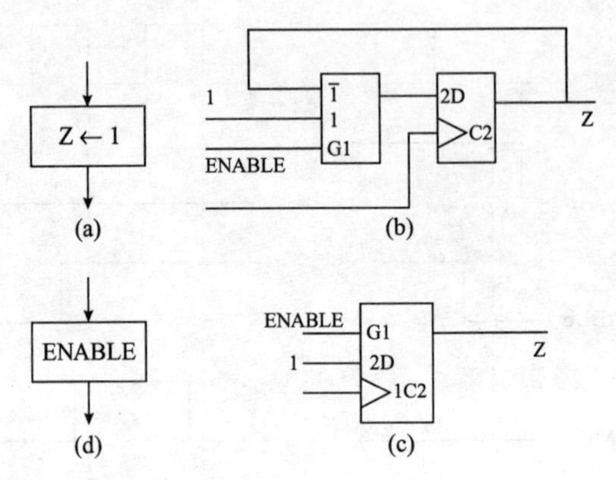

Figure 7.7 Extended ASM chart notation.

Returning to the example of Figure 7.3, it can be seen that the left-hand state machine corresponds to a controller, while the right-hand state machine, the counter, corresponds to the datapath. The TIMED signal is a status signal, as shown in Figure 7.6, while the ENABLE signal is a control signal. We look at a more significant example of datapath/controller partitioning in Section 7.4.

The datapath would normally contain registers. As the functionality of the system is mainly contained in the datapath, the system can be described in terms of *register transfer operations*. These register transfer operations can be described using an extension of ASM chart notation. In the simplest case, a registered output can be indicated, as shown in Figure 7.7(a). This notation means that Z takes the value 1 *at the end* of the state indicated, and *holds that value* until it is reset. If, in this example, Z is reset to 0 and it is only set to 1 in the state shown, the registered output would be implemented as a flip-flop and multiplexer, as shown in Figure 7.7(b), or simply as an enabled flip-flop as shown in Figure 7.7(c). In either implementation, the ENABLE signal is only asserted when the ASM is in the indicated state. Thus, the ASM chart could equally include the ENABLE signal, as shown in Figure 7.7(d).

A more complex example is shown in Figure 7.8. In state 00, three registers, B_0, B_1, and B_2, are loaded with inputs X_0, X_1, and X_2, respectively. Input A then determines whether a shift left, or multiply by 2, is performed ($A = 0$) or a shift right, or divide by 2 ($A = 1$) in the next state. If a divide by 2 is performed, the value of the least significant bit is tested, so as always to round up. From the ASM chart we can derive next state equations for the controller, either formally or by inspection:

$$S_0^+ = \bar{S}_0 \cdot \bar{S}_1 \cdot (\bar{A} + \bar{X}_0)$$
$$S_1^+ = \bar{S}_0 \cdot \bar{S}_1 \cdot A$$

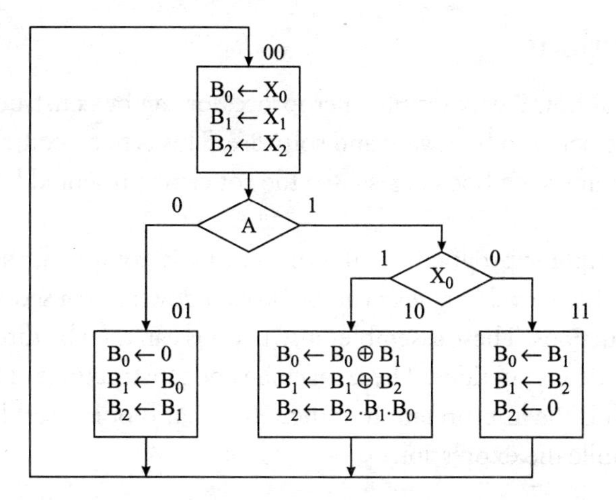

Figure 7.8 ASM chart of a partitioned design.

The datapath part of the design can be implemented using registers for B_0, B_1, and B_2 and multiplexers, controlled by S_0 and S_1, to select the inputs to the registers, as shown in Figure 7.9. It is also possible to implement the input logic using standard gates and thus to simplify the logic slightly.

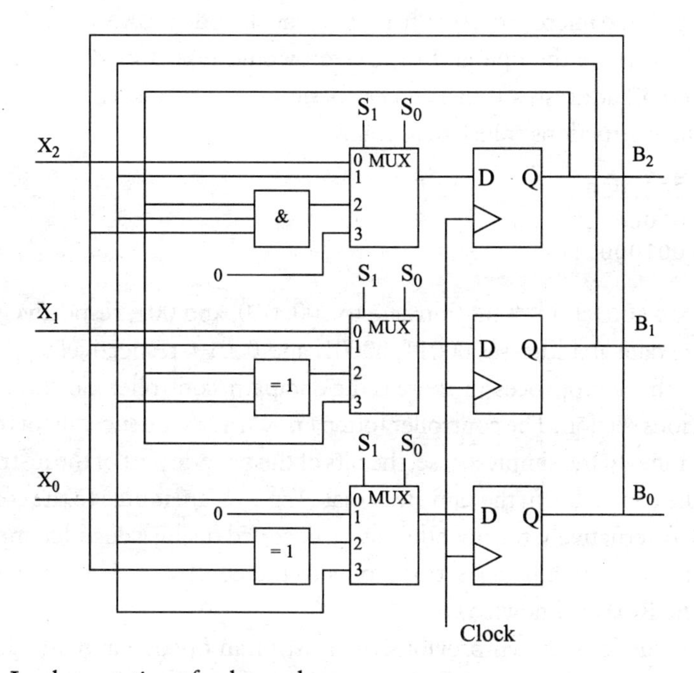

Figure 7.9 Implementation of a datapath.

7.3 Instructions

Before looking at how a very simple microprocessor can be constructed, we examine the interface between hardware and software. This is not a course on computer architecture—many such books exist—so the concepts presented here are deliberately simplified.

When a computer program, written in, say, C, is compiled, the complex expressions of the high-level language can be broken down into a sequence of simple assembler instructions. These assembler instructions can then be directly translated into machine code instructions. These machine code instructions are sets of, say, 8, 16, or 32 bits. It is the interpretation of these bits that is of interest here.

Let us compile the expression

```
a = b + c;
```

to a sequence of assembly code instructions:

```
LOAD b
ADD c
STORE a
```

The exact interpretation of these assembler instructions is explained in the next section. If the microprocessor has 8 bits, the opcode (`LOAD`, `STORE`, etc.) might require 3 bits, while the operand (`a`, `b`, etc.) would take 5 bits. This allows for 8 opcodes and 32 addresses (this is a *very* basic microprocessor). Hence, we might find that the instructions translate as follows.

```
LOAD b   00000001
ADD c    01000010
STORE a  00100011
```

that is, `LOAD`, `ADD`, and `STORE` translate to 000, 010, and 001, respectively, while `a`, `b`, and `c` are data at addresses 00011, 00001, and 00010, respectively.

Within the microprocessor there is the datapath/controller partition described in the previous section. The controller (often known as a sequencer in this context) is a state machine. In the simplest case, the bits of the opcode part of the instruction are inputs to the controller, in the same way that A and X_0 are inputs to the controller of Figure 7.8. Alternatively, the opcode may be decoded (using a decoder implemented in ROM) to generate a larger set of inputs to the controller. The decoder pattern stored in the ROM is known as *microcode*.

The instructions shown previously consist of an opcode and an address. The data to be operated upon must be subsequently obtained from the memory addresses given in the instruction. This is known as *direct* addressing. Other addressing modes are possible. Suppose we wish to compile:

```
a = b + 5;
```

This translates to:

```
LOAD b
ADD 5
STORE a
```

How do we know that the 5 in the ADD instruction means the value "5" and not the data stored at address 5? In assembler language, we would normally use a special notation, for example, "ADD #5," where the "#" indicates to the assembler that the following value is to be interpreted as a value and not as an address. This form of addressing is known as *immediate* mode addressing.

When the microprocessor executes an immediate mode instruction, different parts of the datapath are used compared with those activated by a direct mode instruction. Hence, the controller goes through a different sequence of states, and thus the opcodes for an immediate mode ADD and a direct mode ADD must be different. In other words, from the point of view of the microprocessor, instructions with different addressing modes are treated as totally distinct instructions and have different opcodes.

7.4 A Simple Microprocessor

Using the idea of partitioning a design into a controller and datapath, we now show how a very basic microprocessor can be designed. We want to be able to execute simple direct mode instructions such as those described in the previous section. Let us first consider the components of the datapath that we need. In order to simplify the routing of data around the microprocessor, we assume the existence of a single bus. More advanced designs would have two or three buses, but one bus is sufficient for our needs. For simplicity, we assume that the bus and all the datapath components are 8 bits wide, although we make the SystemVerilog model, in the next section, parameterizable. Because the single bus may be driven by a number of different components, each of those components will use three-state buffers to ensure that only one component is attempting to put valid data on the bus at a time. We keep the design fully synchronous, with a single clock driving all sequential blocks. We also include a single asynchronous reset to initialize all sequential blocks. A block diagram of the microprocessor is shown in Figure 7.10.

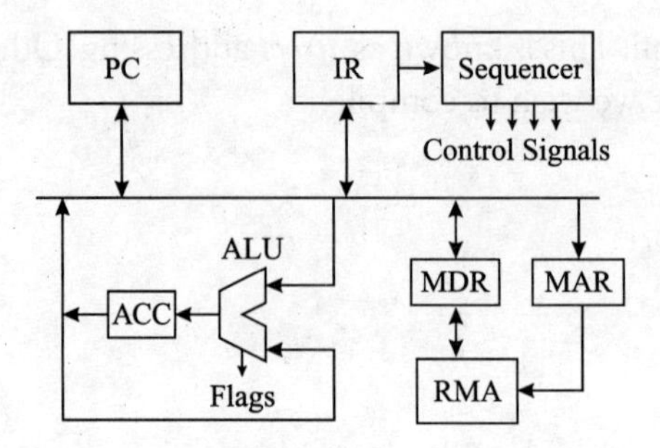

Figure 7.10 Datapath of a central processing unit (CPU).

The program to be executed by the microprocessor will be held in memory together with any data. Memory, such as SRAM, is commonly asynchronous; therefore, synchronous registers will be included as buffers between the memory and the bus for both the address and data signals. These registers are the memory address register (MAR) and memory data register (MDR).

The arithmetic and logic unit (ALU) performs the arithmetic operations (e.g., ADD). The ALU is a combinational block. The result of an arithmetic operation is held in a register, called the accumulator (ACC). The inputs to the ALU are the bus and the ACC. The ALU may also have further outputs, or flags, to indicate that the result in the ACC has a particular characteristic, such as being negative. These flags act as inputs to the sequencer.

The various instructions of a program are held sequentially in memory. Therefore, the address of the next instruction to be executed needs to be stored. This is done using the program counter (PC), which also includes the necessary logic to automatically increment the address held in the PC. If a branch is executed, the program executes out of sequence, so it must also be possible to load a new address into the PC.

Finally, an instruction taken from the memory needs to be stored and acted upon. The instruction register (IR) holds the current instruction. The bits corresponding to the opcode are inputs to the sequencer, which is the state machine controlling the overall functioning of the microprocessor.

The sequencer generates a number of control signals. These determine which components can write to the bus, which registers are loaded from the bus, and which ALU operations are performed. The control signals for this microprocessor are listed in Table 7.1.

Table 7.1 Control Signals of a Microprocessor

Signal	Description
ACC_bus	Drive bus with contents of ACC (enable three-state output)
load_ACC	Load ACC from bus
PC_bus	Drive bus with contents of PC
load_IR	Load IR from bus
load_MAR	Load MAR from bus
MDR_bus	Drive bus with contents of MDR
load_MDR	Load MDR from bus
ALU_ACC	Load ACC with result from ALU
INC_PC	Increment PC and save the result in PC
Addr_bus	Drive bus with operand part of instruction held in IR
CS	Chip select
	Use contents of MAR to set up memory address
R_NW	Read, not write
	When false, contents of MDR are stored in memory
ALU_add	Perform an add operation in the ALU
ALU_sub	Perform a subtract operation in the ALU

Figure 7.11 shows the ASM chart of the microprocessor sequencer. Six clock cycles are required to complete each instruction. The execution cycle can be divided into two parts: the *fetch* phase and the *execute* phase. In the first state of the fetch phase, *s*0, the contents of the PC are loaded, via the bus, into MAR. At the same time the address in the PC is incremented by 1. In state *s*1, the CS and R_NW signals are both asserted to read into MDR the contents of the memory at the address given by MAR. In state *s*2, the contents of MDR are transferred to the IR via the bus.

In the execute phase, the instruction, now held in the IR, is interpreted and executed. In state *s*3, the address part of the instruction, the operand, is copied back to the MAR, in anticipation of using it to load or store further data. If the opcode held in the IR is STORE, control passes through *s*4 and *s*5, in which the contents of ACC are transferred to the MDR, then to be written into memory (at the address previously stored in the MAR) when CS is asserted. If the opcode is not STORE, CS and R_NW are asserted in state *s*6, to read data from memory into the MDR. If the opcode is LOAD, the contents of the MDR are transferred to the ACC in state *s*7; otherwise, an arithmetic operation is performed by the ALU using the data in the ACC and in the MDR in state *s*8. The results of this operation are stored in the ACC.

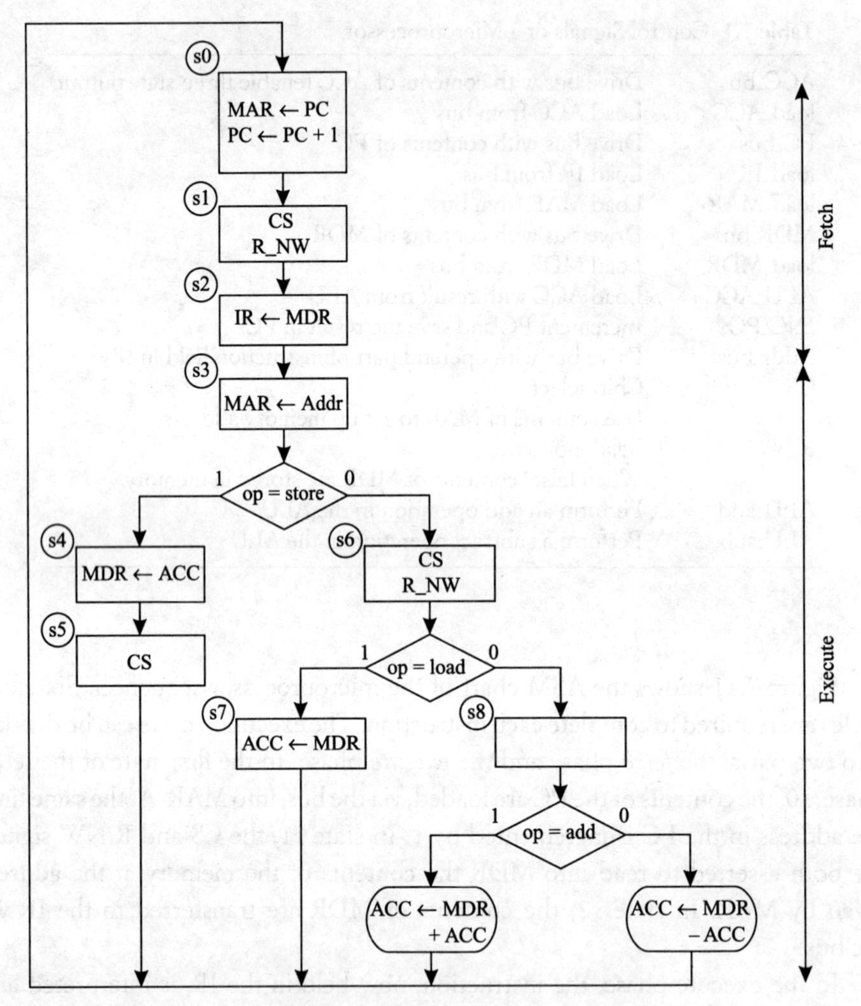

Figure 7.11 ASM chart of a microprocessor.

The ASM chart of Figure 7.11 shows register transfer operations. In Figure 7.12, the ASM chart instead shows the control signals that are asserted in each state. Either form is valid, although that of Figure 7.11 is more abstract.

This processor does not include branching. Hence, it is of little use for running programs. Let us extend the microprocessor to include a branch if the result of an arithmetic operation (stored in the ACC) is not zero (BNE)[1]. The ALU has a

1. Branch if not equal to zero.

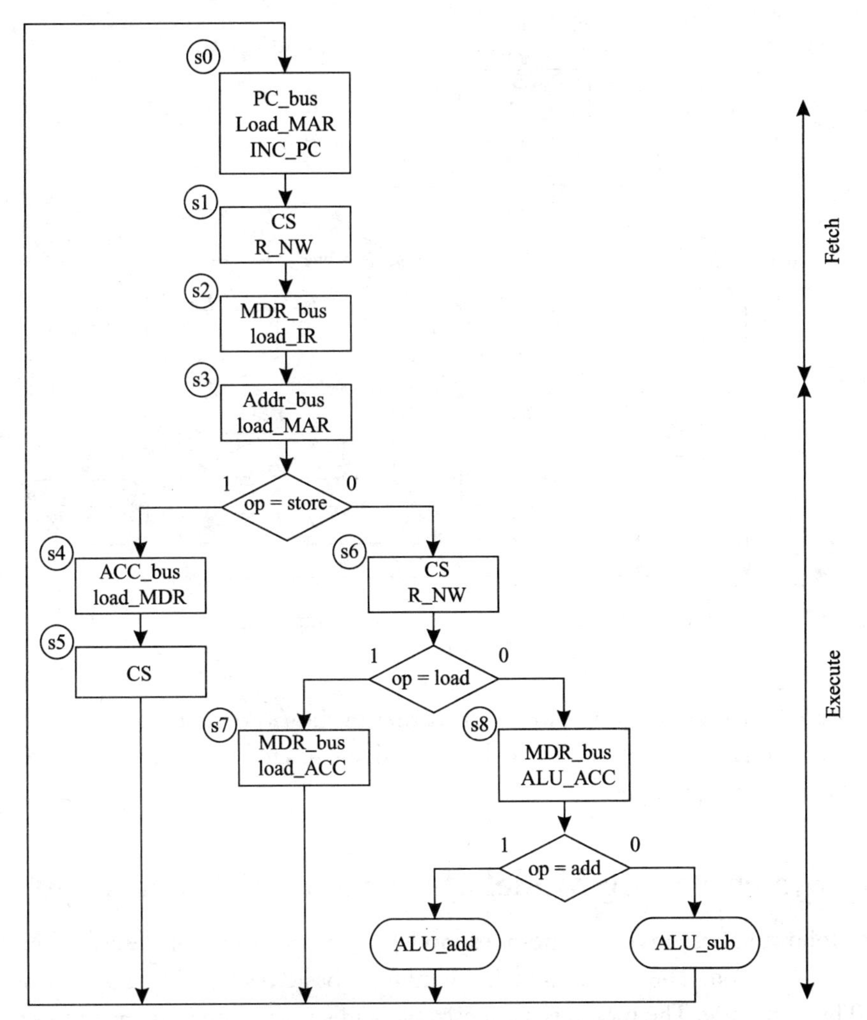

Figure 7.12 Alternative form of the microprocessor ASM chart.

zero flag, which is true if the result it calculates is zero and which is an input to the sequencer. Here, we shall implement this branch instruction in a somewhat unusual manner. All the instructions in this example are direct mode instructions. To implement immediate mode instructions would require significant alteration of the ASM chart. Therefore, we implement a "direct mode branch." The operand of a BNE instruction is not the address to which the microprocessor will branch (if the zero flag is true), but the address at which this destination address is stored.

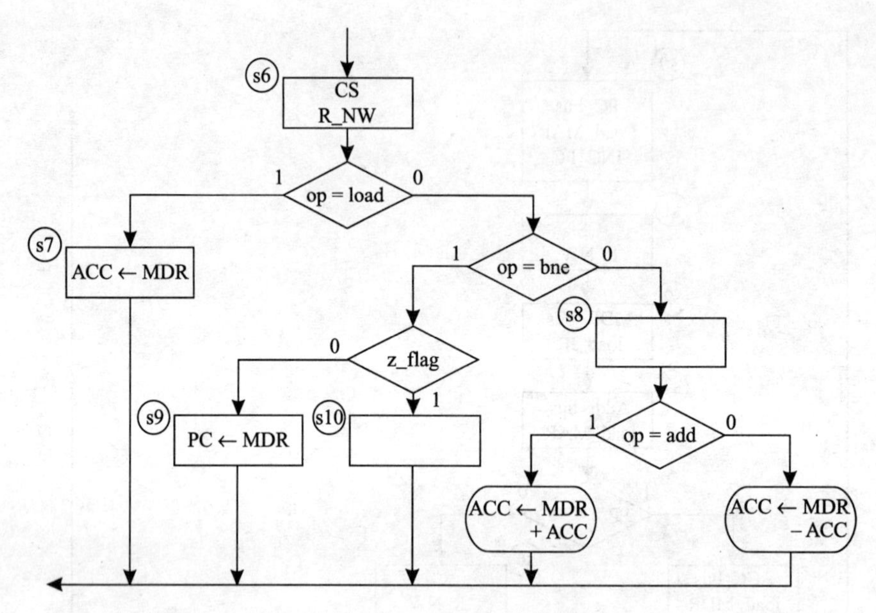

Figure 7.13 Modification of the ASM chart to include branching.

Figure 7.13 shows how the lower right corner of the ASM chart can be modified to include this branch. An additional control signal has to be included: load_PC, to load the PC from the bus.

7.5 SystemVerilog Model of a Simple Microprocessor

The following SystemVerilog modules model the microprocessor described in the previous section. The entire model, including a basic testbench, runs to around 320 lines of code. The model is synthesizable and so could be implemented on an FPGA.

The first file, cpu_defs.v, is a set of definitions contained in a **package**. The definitions are public and may be used in any unit that uses an **import** statement. The opcodes are defined by bit patterns. The size of the bus and the number of bits in the opcode are defined by parameters. The use of this file means that the size of the CPU and the actual opcodes can be changed without altering any other part of the model. This is important to maintain the modularity of the design.

```
package cpu_defs;

parameter WORD_W = 8;
parameter OP_W = 3;
```

```
enum logic[2:0] {LOAD=3'b000,
                 STORE=3'b001,
                 ADD=3'b010,
                 SUB=3'b011,
                 BNE=3'b100} opcodes;

endpackage
```

The modules share a common bus and are linked by a number of control signals. In this example, we gather together all these signals into an **interface**. While this is not really necessary for a design of this size, it illustrates a useful construct, with little extra code. The basic idea is that an interface is instantiated in the same way as any other module. The signals that pass between modules are declared in the interface. The particular signals connecting to a specific module are declared in a named **modport**. As we will see in the following, this name is used within each of the modules. In this example, the clock, reset, and main system bus are declared as external signals to the interface. This is done because the clock and reset signals are generated externally, and if the bus were totally internal—in other words, if there were no outputs from the system—a synthesis tool would conclude that the system could be optimized away to nothing!

```
import cpu_defs::*;

interface CPU_bus (input logic clock, n_reset,
                   inout wire [WORD_W-1:0] sysbus);

logic ACC_bus, load_ACC, PC_bus, load_PC, load_IR,
      load_MAR, MDR_bus, load_MDR, ALU_ACC, ALU_add,
      ALU_sub, INC_PC, Addr_bus, CS, R_NW, z_flag;

logic [OP_W-1:0] op;

modport IR_port(input clock, n_reset, Addr_bus,
                      load_IR,
                inout sysbus,
                output op);

modport RAM_port (input clock, n_reset, MDR_bus,
                        load_MDR, load_MAR, CS, R_NW,
                  inout sysbus);

modport ALU_port (input clock, n_reset, ACC_bus,
                        load_ACC, ALU_ACC, ALU_add, ALU_sub,
                  inout sysbus,
                  output z_flag);
```

```
modport PC_port (input clock, n_reset, PC_bus, load_PC,
                       INC_PC,
                 inout sysbus);

modport seq_port (input clock, n_reset, z_flag,
                  input op,
                  output ACC_bus, load_ACC, PC_bus,
                         load_PC, load_IR, load_MAR,
                         MDR_bus, load_MDR, ALU_ACC,
                         ALU_add, ALU_sub, INC_PC,
                         Addr_bus, CS, R_NW);

endinterface
```

The controller or sequencer is described by the ASM chart in Figures 7.12 and 7.13. The SystemVerilog description therefore also takes the form of a state machine. The inputs to the state machine are the clock, reset, an opcode, and the zero flag from the accumulator. The outputs are the control signals of Table 7.1. All of these signals are routed through the **interface**. The **modport** from the **interface** is referenced in the module header. Thus, bus is declared to be of type CPU_bus.seq_port. All external signals coming through this interface are prefixed with bus. The advantage of the **interface** can now be seen. If a control signal is changed, the change is made in the interface declaration and in the module bodies concerned—we don't have to worry about changing the module headers. Notice that a two-block model is used. Notice, too, that all the output signals are given a default value at the start of the next state and output logic block. At the end of the case statement in the combinational block, a default statement assigns x values to the state variable. This will be treated as don't care values in synthesis, and would highlight any error during simulation.

```
import cpu_defs::*;

module sequencer (CPU_bus.seq_port bus);

enum {s0, s1, s2, s3, s4, s5, s6, s7, s8, s9, s10}
  state;

always_ff @(posedge bus.clock, negedge bus.n_reset)
  begin: seq
    if (!bus.n_reset)
      state <= s0;
    else
```

```
      case (state)
         s0: state <= s1;
         s1: state <= s2;
         s2: state <= s3;
         s3: if (bus.op == STORE)
                state <= s4;
             else
                state <= s6;
         s4: state <= s5;
         s5: state <= s0;
         s6: if (bus.op == LOAD)
                state <= s7;
             else if (bus.op == BNE)
                if (~bus.z_flag)
                   state <= s9;
                else
                   state <= s10;
             else
                   state <= s8;
         s7: state <= s0;
         s8: state <= s0;
         s9: state <= s0;
         s10: state <= s0;
   endcase
 end

always_comb
 begin: com
 // reset all the control signals to default
 bus.ACC_bus = '0;
 bus.load_ACC = '0;
 bus.PC_bus = '0;
 bus.load_PC = '0;
 bus.load_IR = '0;
 bus.load_MAR = '0;
 bus.MDR_bus = '0;
 bus.load_MDR = '0;
 bus.ALU_ACC = '0;
 bus.ALU_add = '0;
 bus.ALU_sub = '0;
 bus.INC_PC = '0;
 bus.Addr_bus = '0;
 bus.CS = '0;
 bus.R_NW = '0;
 case (state)
   s0: begin
```

```
            bus.PC_bus = '1;
            bus.load_MAR = '1;
            bus.INC_PC = '1;
            bus.load_PC = '1;
            end
      s1: begin
            bus.CS = '1;
            bus.R_NW = '1;
            end
      s2: begin
            bus.MDR_bus = '1;
            bus.load_IR = '1;
            end
      s3: begin
            bus.Addr_bus = '1;
            bus.load_MAR = '1;
            end
      s4: begin
            bus.ACC_bus = '1;
            bus.load_MDR = '1;
            end
      s5: begin
            bus.CS = '1;
            end
      s6: begin
            bus.CS = '1;
            bus.R_NW = '1;
            end
     s7: begin
           bus.MDR_bus = '1;
           bus.load_ACC = '1;
           end
     s8: begin
           bus.MDR_bus = '1;
           bus.ALU_ACC = '1;
           bus.load_ACC = '1;
           if (bus.op == ADD)
             bus.ALU_add = '1;
           else if (bus.op == SUB)
             bus.ALU_sub = '1;
           end
     s9: begin
           bus.MDR_bus = '1;
           bus.load_PC = '1;
           end
    s10: ;
```

```
  endcase
  end
endmodule
```

The datapath side of the design, as shown in Figure 7.10, has been described in four parts. Each of these parts is similar to the type of sequential building block described in Chapter 5. The system bus is described as a bidirectional port in each of the following four modules. An assignment sets a high impedance state onto the bus unless the appropriate output enable signal is set. Notice the use of the replication operator. The first module models the ALU and the ACC.

```
import cpu_defs::*;

module ALU (CPU_bus.ALU_port bus);

logic [WORD_W-1:0] acc;

assign bus.sysbus = bus.ACC_bus ? acc : 'z;
assign bus.z_flag = acc == 0 ? '1 : '0;

always_ff @(posedge bus.clock, negedge bus.n_reset)
  begin
  if (!bus.n_reset)
    acc <= 0;
  else
    if (bus.load_ACC)
      if (bus.ALU_ACC)
        begin
        if (bus.ALU_add)
          acc <= acc + bus.sysbus;
        else if (bus.ALU_sub)
          acc <= acc - bus.sysbus;
        end
      else
        acc <= bus.sysbus;
  end
endmodule
```

The program counter is similar in structure to the ALU and the ACC. Notice the use of replication and concatenation to set the most significant bits on the system bus.

```
import cpu_defs::*;

module PC (CPU_bus.PC_port bus);

logic [WORD_W-OP_W-1:0] count;
```

```
assign bus.sysbus = bus.PC_bus ?
       {{OP_W{1'b0}},count} : 'z;

always_ff @(posedge bus.clock, negedge bus.n_reset)
  begin
  if (!bus.n_reset)
    count <= 0;
  else
    if (bus.load_PC)
      if (bus.INC_PC)
        count <= count + 1;
      else
        count <= bus.sysbus;
  end
endmodule
```

The instruction register is basically an enabled register.

```
import cpu_defs::*;

module IR (CPU_bus.IR_port bus);

logic  [WORD_W-1:0] instr_reg;

assign bus.sysbus = bus.Addr_bus ?
{{OP_W{1'b0}},instr_reg[WORD_W-OP_W-1:0]} :
'z;

always_comb
  bus.op = instr_reg[WORD_W-1:WORD_W-OP_W];

always_ff @(posedge bus.clock, negedge bus.n_reset)
  begin
  if (!bus.n_reset)
    instr_reg <= 0;
  else
    if (bus.load_IR)
      instr_reg <= bus.sysbus;
  end
endmodule
```

The memory module is, again, very similar to the static RAM in Chapter 5. A short program has been loaded in the RAM. In order to make the model parameterizable, the "program" has been written as an opcode followed by some 0s followed by an address. Because we know the size of the address (3 bits in each case), we can set the number of zeros by replication. The replication operator is written as {N{1'b0}} to specify, for example, N instances of a single 0. Because the replication number is an expression, it is written in parentheses.

```
import cpu_defs::*;

module RAM (CPU_bus.RAM_port bus);

logic [WORD_W-1:0] mdr;
logic [WORD_W-OP_W-1:0] mar;
logic [WORD_W-1:0] mem [0:(1<<(WORD_W-OP_W))-1];
int i;

assign bus.sysbus = bus.MDR_bus ? mdr : 'z;

always_ff @(posedge bus.clock, negedge bus.n_reset)
  begin
  if (!bus.n_reset)
    begin
    mdr <= 0;
    mar <= 0;
    mem[0] <= {LOAD, {(WORD_W-OP_W-3){1'b0}},3'd4};
    mem[1] <= {ADD, {(WORD_W-OP_W-3){1'b0}},3'd5};
    mem[2] <= {STORE,{(WORD_W-OP_W-3){1'b0}},3'd6};
    mem[3] <= {BNE, {(WORD_W-OP_W-3){1'b0}},3'd7};
    mem[4] <= 2;
    mem[5] <= 2;
    mem[6] <= 0;
    mem[7] <= 0;
    for (i = 8; i < (1<<(WORD_W-OP_W)); i+=1)
      mem[i] <= 0;
    end
  else
    if (bus.load_MAR)
      mar <= bus.sysbus[WORD_W-OP_W-1:0];
    else if (bus.load_MDR)
      mdr <= bus.sysbus;
    else if (bus.CS)
      if (bus.R_NW)
        mdr <= mem[mar];
      else
        mem[mar] <= mdr;
  end
endmodule
```

The various parts of the microprocessor can now be pulled together by instantiating them. Here, we use the SystemVerilog default style of argument passing. In this style, we assume that arguments have the same internal and external names, and this allows us to simply write (.*).

```
import cpu_defs::*;

module CPU (input logic clock, n_reset,
            inout wire [WORD_W-1:0] sysbus);

CPU_bus bus (.*);

sequencer s1  (.*);

IR i1  (.*);

PC p1 (.*);

ALU a1 (.*);

RAM r1 (.*);

endmodule
```

The following piece of Verilog generates a clock and reset signal to allow the program defined in the RAM module to be executed. Obviously, this testbench would not be synthesized.

```
import cpu_defs::*;

module TestCPU;

logic  clock, n_reset;
wire [WORD_W-1:0] sysbus;

CPU c1 (.*);

always
  begin
#10ns clock = 1'b1;
#10ns clock = 1'b0;
end

initial
begin
n_reset = 1'b1;
#1ns n_reset = 1'b0;
#2ns n_reset = 1'b1;
end

endmodule
```

Summary

In this chapter we looked at linked ASM charts and splitting a design between a controller, which is designed using formal sequential design methods, and a datapath that consists of standard building blocks. The example of a simple CPU has been used to illustrate this partitioning. The SystemVerilog model can be both simulated and synthesized.

Further Reading

Formal techniques exist for partitioning state machines. These are described in Unger [23]. The controller/datapath model is used in a number of high-level synthesis tools; see, for example, de Micheli [6]. The CPU model is based on an example in Maccabe [13].

Exercises

7.1 Any synchronous sequential system can be described by a single ASM chart. Why then might it be desirable to partition a design? Describe a general partitioning scheme.

7.2 A counter is to be used as a delay for a simple controller, to generate a ready signal, 10 clock cycles after a start signal has been asserted. Show how the interaction between the controller and the counter can be represented in ASM chart notation.

7.3 A microprocessor has a number of addressing modes. Describe the immediate and direct addressing modes.

7.4 What structures are needed in a microprocessor to implement a "branch if negative" instruction? Describe the register transfer operations that would occur in the execution of such an instruction and show the sequence of events on a timing diagram.

7.5 The ASM chart of Figures 7.11 and 7.13 implements a branch instruction with a direct mode operand. Modify the ASM chart to show how the microprocessor could branch to an address given by an immediate mode operand.

7.6 Modify the SystemVerilog model of the microprocessor to implement an immediate mode "branch if not equal to zero."

8 Writing Testbenches

Writing a synthesizable model of a piece of hardware is only half (or perhaps less than half) of the design problem. It is essential to know that the model does the task for which it is intended. It would, of course, be possible to do this the hard way—by synthesizing the hardware and testing the design in the final context in which it is to be used. This could be a very expensive and dangerous approach.

The alternative is to verify the hardware before synthesis. In practice, this means that the hardware has to be simulated. In order to simulate a SystemVerilog model, stimuli have to be applied to the model and the responses of the model have to be analyzed. For portability and to avoid having to learn a new set of language constructs, the stimuli and response analysis routines are written in SystemVerilog. It is tempting to argue that with FPGAs, it can be as fast to make changes to the hardware as it is to simulate. There is some truth to this, inasmuch as the quality of the verification cannot be truly known until the actual hardware is tested, but simulation should always be used to check any changes before synthesis is done.

We use the term *testbench* to describe a piece of SystemVerilog written to verify a synthesizable model. There are two basic features of a testbench that distinguish it from a synthesizable model. First, a testbench has no inputs or outputs; it is the entire world as far as the model is concerned. In a simulation, we can have access to every part of a model; therefore, this lack of input and outputs does not restrict us in any way. Second, because a testbench is *never* synthesized, we can use the entire SystemVerilog language. This freedom to use the entire language can present its own difficulties. By sticking to an agreed upon subset of SystemVerilog it is straightforward to write portable, synthesizable hardware models. Because of the definition of the SystemVerilog simulation cycle (see Chapter 9), it cannot be guaranteed that an arbitrary piece of SystemVerilog code will execute in exactly the same way on two simulators. Therefore, it is very easy to write testbenches that behave differently, and that give different simulation results, on different simulators.

We already introduced several examples of testbenches in earlier chapters. We start here by recapping the examples from earlier chapters. Just as synthesizable models should be written in a modular manner to allow reuse, testbenches should also be written as modules as will be shown. The final two sections of this chapter discuss constrained random test generation and assertion-based verification. Both of these concepts have a considerable amount of new programming constructs associated with them, so only some introductory examples will be given.

8.1 Basic Testbenches

In Chapter 3, a basic testbench for a two-input AND gate was given.

```
module TestAnd2;

  wire a,b,c;

  And2 g1 (c, a, b);

initial
  begin
  a = '0;
  b = '0;
  #100ps a = '1;
  #50ps b = '1;
  end

endmodule
```

The testbench has an **initial** procedure containing a sequence of assignments to a and b. The **initial** keyword does not mean that the procedure initializes signals, but rather that the procedure is only executed once. Notice that the delay value is a relative delay and appears on the left side of a blocking assignment. This style is suitable for testbenches, but should not be used for modeling synthesizable hardware.

8.1.1 Clock Generation

The most important signal in any sequential design is the clock. In the simplest case, a clock can be generated by inverting its value at a regular interval.

The default value of any variable or net is X. Simply inverting a signal at a regular interval will invert the X value. Therefore, the signal has to be initialized. An example of this was first given in Chapter 5.

```
initial
  begin
  clock = '0;
  forever #10ps clock = ~clock;
  end
```

8.1.2 Reset and Other Deterministic Signals

Also in Chapter 5, there was an example of a reset:

```
initial
  begin
       n_reset = '1;
  #1ns n_reset = '0;
  #1ns n_reset = '1;
  end
```

We also showed how other non-repeating stimuli could be generated in a similar way. In Chapter 6, there were examples showing how inputs can be synchronized to a clock signal.

8.1.3 Monitoring Responses

While it is possible to determine whether a simple design is correct by observing the waveforms generated by a simulator, this approach soon becomes impractical for large designs simulated over thousands of clock cycles. In Chapter 5, the system calls **$write**, **$display**, **$monitor**, and **$strobe** were introduced. So, for example, the state of a variable at the end of the simulation cycle at a given time can be written using a command like:

```
$strobe("%t Counter has value %d", $time, count);
```

8.1.4 Dumping Responses

The results of a simulation can be "dumped" to a file for later analysis or for display in a waveform viewer. (But note that this is not usually necessary if the viewer is built into the simulator.) The name of the file is specified using the `$dumpfile` system call and the variables of the system saved using the `$dumpvars` call. So, for example, we could write:

```
initial
  begin
  $dumpfile("results.vcd");
  $dumpvars;
  end
```

`$dumpvars` saves all variables in this case, but arguments can be used to specify a subset to save.

8.1.5 Test Vectors from a File

It is sometimes convenient to keep a set of test vectors in a separate text file. For example, suppose that the following set of inputs were to be applied to a device.

```
0000
0010
1001
0101
1111
```

If these vectors are stored in a text file `vectors.txt`, they could be read into an array and applied to a device under test as shown in the following code fragment.

```
logic [3:0] test_vector [0:4];
logic [3:0] a;

initial
  begin
  $readmemb("vectors.txt", test_vector);
  for (int i = 0; i <= 4; i++)
    a = test_vector[i];
  end
```

The variable a would be the input to the device under test.

8.2 Testbench Structure

In the examples seen thus far, the structure of the testbench is that of the first example of this chapter. The testbench consists of a single module, within which the circuit under test (CUT)—or, more accurately, device under verification (DUV)—is instantiated. The stimulus generation and response checking is contained within that testbench in a number of procedures. This structure is fine, if you only want to

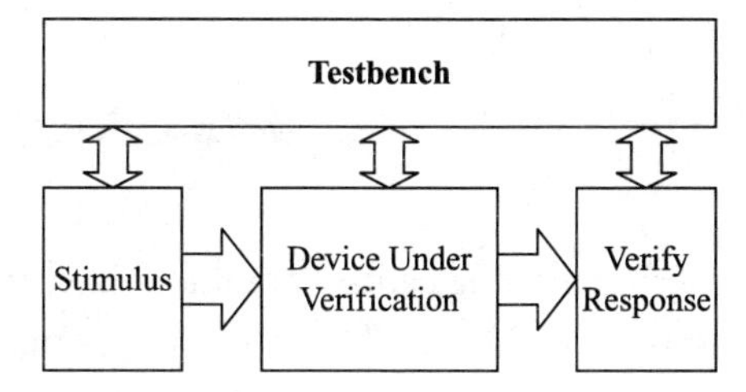

Figure 8.1 Modular testbench structure.

run one simulation or if the parts of the testbench will be used only once. If multiple simulations are needed to test different parts of a design, it makes sense to adopt a modular structure.

The basic idea is illustrated in Figure 8.1. The stimulus and verification sections are instantiated within separate units, just like the DUV. The clock and reset are special cases and are generated separately, either within the top-level testbench or in a separate module.

As an example, to illustrate the concepts in the rest of this chapter, let us consider the sequential Booth multiplier of Section 5.7. The multiplier takes two twos-complement numbers and calculates the product after a certain number of clock cycles. It is not required that the multiplier and the multiplicand have the same number of bits, so as well as verifying that the circuit gives the correct answer, we would also like to know that it generates the correct answer after the correct number of cycles for inputs of different bit widths. This suggests that more than one testbench may be required. It is, therefore, common to speak of more than one *testcase*, rather than of a single *testbench*.

The top-level testbench, corresponding to the structure of Figure 8.1, follows. The testbench simply declares parameters and signals, and the four units corresponding to the clock generator, stimulus generator, device under verification, and response verifier are instantiated.

```
module testbooth;

parameter N = 16;

logic signed [N-1:0] ain, bin;
logic signed [2*N-1:0] qout;
logic clk, load, n_reset;
logic ready;

clock_gen #(10.0, 10.0) c0 (.*);
stimulus s0 (.*);
```

```
booth #(N, N, 2*N) b0 (.*);
verify v0 (.*);

endmodule
```

The clock (and reset) generator can be parameterized to make it usable as a general block in any testbench. Notice that this is very similar to the asymmetric clock given in Chapter 5 and so will not be explained further here.

```
module clock_gen #(parameter ClockFreq_MHz = 100.0,
                             ResetWidth=10.0)
       (output logic clk, n_reset);
timeunit 1ns;
timeprecision 100ps;

parameter ClockHigh = (500.0)/ClockFreq_MHz;

initial
  begin
        n_reset = '1;
        clk = '0;
  #ResetWidth n_reset = '0;
  #ResetWidth n_reset = '1;
  forever #ClockHigh clk = ~clk;
  end

endmodule
```

8.2.1 Programs

In all the design examples given in previous chapters, sequential logic (in **always_ff** procedures) is always modeled using *nonblocking* assignments (<=); combinational logic (in **always_comb** procedures) is always modeled using *blocking* assignments (=). The reason for this will be explained in full in Chapter 9, but it is sufficient to note for the moment that this convention ensures that, at a clock edge, all combinational logic is updated before any flip-flop outputs change, and signal changes do not ripple through multiple flip-flops on one clock edge.

A similar problem can arise in a testbench. Signals assigned using blocking assignments would happen at the same time as combinational logic and, worse, can happen in any order or can be interleaved with combinational logic changes. Nonblocking assignments with delays can suffer from the same problems. It would be preferable to move all assignments in testbenches into a separate grouping. The **program** block does this. Again, the mechanism is described in Chapter 9.

A **program** block can contain type, variable, and subprogram declarations, together with one or more **initial** blocks. It cannot contain any **always** blocks (or the RTL variants), nor any other block instantiations. Thus, a **program** is very much like a conventional software program—it executes one or more threads and then terminates.

A simple program block to generate a stimulus for the multiplier is given in the following. The two inputs, ain and bin, are assigned values after the reset phase and then load is asserted for one clock edge.

```
program stimulus #(parameter NA = 16, NB = 16)
        (output logic signed [NA-1:0] ain,
         output logic signed [NB-1:0] bin,
         output logic load);

initial
  begin
  #30ns ain = -10;
        bin = 10;
  #10ns load = '1;
  #10ns load = '0;
  end

endprogram
```

Of course, it would be preferable to parameterize the timing of this block. This is left as an exercise.

It is now relatively easy to build up a collection of test cases, in which different stimulus blocks are instantiated within different top-level modules. For a larger example, different stimulus blocks can be combined in different ways to create different tests. Alternatively, generic stimulus blocks can be created, for which particular test cases can be created using different top-level modules.

A few further points should be made about program blocks. First, it is possible to generate inputs from multiple program blocks. As with hardware, a variable or net should not be assigned values from two or more programs, or contention may arise. Second, it is also possible to call the same program more than once (just like a real subroutine), but with different parameters. By default, a program—or any SystemVerilog block—has static storage. Multiple uses of the same program will confuse matters. It is therefore desirable to declare programs with the **automatic** keyword:

```
program automatic stimulus (...);
```

Finally, the clock generator was declared using a **module**, not a **program**. The clock *is* different from other inputs (as is a power-on, asynchronous reset).

There is sense in asserting it at the same time as any combinational logic because it has no effect on that logic and because it needs to be asserted before any nonblocking assignments in the flip-flops. In effect, think of the clock generator as a piece of hardware.

8.3 Constrained Random Stimulus Generation

One of the greatest difficulties in writing testbenches is knowing what stimuli to apply. One approach is to use the "corner" cases. For example, to test a multiplier, combinations of 0, the largest positive number, and the largest negative number might be applied. If the multiplier gives the correct answers for these values, it ought to give correct answers for any number. To be on the safe side, we would probably include some other random values. What do we mean by *random*? If we had to choose some numbers "randomly," we would probably chose values such as 10 or 100, because the product is easy to check. So human-chosen values would probably not really be random. Equally, we might consciously or otherwise choose values that we would expect to see. Choosing random values automatically would give us a wider range of data (that we might be too lazy to generate ourselves) and might give us combinations that reveal something unexpected about our design.

Completely random data might, however, include absurd combinations. For example, testing a microprocessor by randomly choosing both opcodes *and* data might give situations that could never arise. Thus, *constrained* random test generation has attracted wide attention. SystemVerilog includes constructs to allow constrained random test generation, but a full description of the topic would take an entire book (see Further Reading). Here, we will give a brief introduction to the topic.

8.3.1 Object-Oriented Programming

Object-oriented programming (OOP) is a software development methodology that, at first sight, has little to do with an HDL such as SystemVerilog. In "conventional" programming languages (C, FORTRAN), programs operate upon data structures. Procedural blocks in SystemVerilog (**`initial`**, **`always`**) work in a very similar manner. OOP turns this model inside out. Data structures and the functions that operate on them are declared together (in *classes*) and programs are assembled by instantiating these classes as *objects*. It is not easy to think of hardware in terms of classes, although SystemC is a hardware description language, based upon C++, a widely used OOP. It should, however, be noted that SystemC is best used for levels of abstraction higher than RTL.

SystemVerilog includes a number of OOP features that can be used for testbench design. As with any programming language, it is possible to do the things described here in different ways, but it is much easier to use these structures.

The basic OOP feature is the *class*. A very simple example of a class in SystemVerilog is:

```
class Twobits;
  bit a, b;
endclass
```

This is somewhat uninteresting! A better example includes one or more functions (or *methods* in OOP terminology).

```
class Twobits;
  bit a, b;

  function implies;
    implies = (~a | b);
  endfunction
endclass
```

The class has to be instantiated as an object and then allocated. Objects are dynamic—they can be created and destroyed. This is one reason why they are not a natural way to model hardware.

```
initial
  begin
  Twobits tb;   // Declare a "handle"
  tb = new();   // And allocate the memory
  tb.a = '0;    // Assign a value to a
  tb.b = '1;
  $display("a => b is %b", tb.implies());
  end
```

Notice the use of the "dot" notation. `tb.a` means variable `a` inside object `tb`. The same notation applies to any method declared inside the class. If more than one object of a given class is declared (each with its own handle), each has to have memory allocated with its own call to `new`.

A lot more can be done with classes: They can be nested and new classes can be derived by inheriting the details of a *base* class. The way in which classes are used in constrained random test generation is a little different from normal OOP practice, however.

8.3.2 Randomization

Instead of assigning values to variables in classes, we want random values to be generated automatically. To do this, we declare the variables as **rand** or **randc**. The difference is that **rand** variables can take any value at any time (subject to constraints), while **randc** variables cycle through all possible values before repeating any value.

Returning to the multiplier example, we can declare a class that contains two variables:

```
class Mults;
  rand logic signed [NA-1:0] ar;
  rand logic signed [NB-1:0] br;
endclass
```

As with the earlier OOP example, the class has to be instantiated as an object and allocated. The variables declared as **rand** are then randomized and assigned to the output variables.

```
initial
  begin
  Mults m;
  m = new();
  m.randomize();
  #30ns ain = m.ar;
        bin = m.br;
  #10ns load = '1;
  #10ns  load = '0;
  end
```

As it stands, this example is not particularly interesting. Only one pair of random values is generated and every simulation will be the same because the random number generator is really only a pseudo-random number generator that uses the same seed each time. (See Section 5.5.3 for another example of a pseudo-random number generator.) We could create a test sequence of, say, 50 random pairs. Notice that the procedure waits for the ready signal from the multiplier to become true before trying to load a new value.

```
initial
  begin
  Mults m;
  m = new();
  repeat (50)
    begin
    m.randomize();
    #30ns ain = m.ar;
```

```
         bin = m.br;
    #10ns load = '1;
    #10ns  load = '0;
    wait (ready);
    end
  end
```

This example chooses *unconstrained* random values for `ar` and `br`. These random values can lie anywhere in the possible range. Clearly, checking the product values will require some further calculations. Suppose, therefore, that we want to limit the range of values such that we can check the answers in our heads. We can apply a **`constraint`** to the values that can be generated. For example, the ranges of `ar` and `br` could be limited to ±20.

```
class Mults;
  rand logic signed [NA-1:0] ar;
  rand logic signed [NB-1:0] br;
  constraint c_range {ar > -20;
                      ar < 20;
                      br > -20;
                      br < 20;};
endclass
```

A more realistic example is based on the microprocessor of Chapter 7. Suppose that we wish to test the microprocessor by using a sequence of random instructions, with random addresses. In a real system, instructions and data are not normally mixed up in the memory. So let us assume that the test will emulate a real system such that the "program" is in the bottom half of the memory and the data is in the top half. For simplicity, our microprocessor has five opcodes and has 8 bits for each operand.

We can declare the opcodes as an enumerated type.

```
typedef enum {LOAD, STORE, ADD, SUB, BNE} opcodes;
```

(Notice that this is not quite the same as the example in Chapter 7. Here, `opcodes` is a type; in the previous example, it was a variable.) The opcode and operand are declared as random variables in a class.

```
class Stimulus;
  rand opcodes op;
  rand logic [7:0] operand; // 0 to 255
endclass
```

We now need to write a constraint that says that if the opcode references part of the program, the operand is in the lower half of the address space, but if the opcode references data, the operand is in the upper half. The only opcode that references another part of the program is the BNE—branch if not equal to zero to another part of the program. The other opcodes all reference data. The constraint is therefore written to check the opcode and to constrain the operand accordingly.

```
constraint c_op {
  if (op == BNE)
    operand <= 127;
  else
    operand > 128;
}
```

The constraint is written, of course, as part of the class declaration. To use this to verify the microprocessor model, that model would have to be restructured. The stimulus generator would replace the RAM block. See Exercise 8.1.

There are several ways of expressing the constraint shown previously. For more information, see the recommendations in the Further Reading section.

8.4 Assertion-Based Verification

In the testbench structure of Figure 8.1, the right-hand box is labeled "Verify Response." In verifying a response, it is implicit that we know what the correct response should be. It is implicit in the discussion in earlier chapters that the functionality of a circuit can be verified by inspecting the waveforms produced by a simulation. It becomes clear very quickly that such an approach is inadequate. It is not really practical to view hundreds of waveforms on a screen and to check that a particular signal changes at exactly the right time, perhaps after thousands of clock cycles and hundreds of input changes.

The **$monitor** or **$display** commands can be used to indicate in a textual form what the response of the circuit should be, but ideally we want to know when an output is not what we expect. It would be possible to tabulate a set of inputs and outputs and to check each output, but for a sequential system, such as the multiplier used to illustrate earlier sections of this chapter, the output might only be valid some clock cycles after the input changes.

All this suggests that what we are really trying to achieve is a comparison between a hardware model and some more abstract model that can be written in a more concise form. Of course, SystemVerilog has the program structures to allow abstract models to be written. On the other hand, *assertions* have been introduced into SystemVerilog to allow sequential behavior to be described in a more abstract manner than standard RTL.

The idea of an assertion is very simple. We state that we believe that something should be true, and if it is not true, an error message is printed. In SystemVerilog, assertions can take two forms: *immediate* and *concurrent* assertions. Immediate assertions are very simple and could be easily written as **if**, **else** statements. Concurrent assertions are much more powerful and will form the main part of this discussion. As with other aspects of the language, there is insufficient space here to go into every detail, and other sources of information are given in the Further Reading section.

To illustrate how assertions can be written, we will use the vending machine example from Chapter 6. We will also show how the outputs of the multiplier can be verified. It may seem that some of the assertions given here are trivially obvious from the code they are designed to verify. Ideally, verification should be performed by a different engineer than the designer. Both should be working independently from a specification. Also, bear in mind that assertions need to be debugged just as much as RTL code, so verification is really the writing of two versions of the same function and comparing them.

The vending machine has four outputs. From the ASM chart (this is the specification against which we are testing the RTL implementation), we can see, for example, that `dispense` and `ready` are never true at the same time. This fact can be stated as an *immediate* assertion in an always procedure. (This could be equivalently written as an **always_comb** procedure, but to avoid confusion, we will only use **always_comb** for synthesizable RTL code.)

```
always @*
  assert (~(dispense && ready))
  else $error("dispense and ready both lit!");
```

This assertion is tested whenever `dispense` or `ready` changes. It would, however, be better to check that this condition is true at a clock edge because we are dealing with a synchronous system. First we write the condition as a **property**.

```
property NotDispenseAndReady;
  @(posedge clock) (~(dispense && ready));
endproperty
```

This property is then tested as part of a *concurrent* assertion.

```
assert property (NotDispenseAndReady);
```

The property and the assertion can be combined, but as a general style guide, it is better to keep them separate to allow properties to be reused.

In general, concurrent assertions are active over a period of time. Temporal properties take two forms: implications and sequences. For example, if the vending machine has just dispensed a ticket or has returned bills, the ready light is on.

If a bill is inserted, we expect that in the *next* clock cycle, the bill light will be lit, indicating that further bills have to be inserted. This can be expressed as a property.

```
property Bill;
 @(posedge clock) ready && (twenty || ten) |=> bill;
endproperty
```

The symbol |=> is a *non-overlapping implication*. In other words, the condition on the left implies that in the next clock cycle, the condition on the right is expected to become true. This property (and those that follow) can be used in **assert** statements, as above.

Instead of putting the clock edge in every property, it is possible to define a default clocking block. The `@(posedge clock)` clause can then be omitted from properties and assertions.

```
default clocking clock_block
 @(posedge clock);
endclocking
```

An *overlapping implication* is one in which the condition on the left implies that the condition on the right is true *in the same clock cycle*. For example, if the vending machine is waiting for a bill, it is not dispensing a ticket at the same time.

```
property BillNotDispense;
  bill |-> !dispense;
endproperty
```

The implication operator means that the property *fails* if bill is true and dispense is also true. The property *passes* if bill is true and dispense is false. The property also passes if bill is false. This is known as a *vacuous* pass. In other words, the property passes because it is never really tested. This is a potential source of false optimism. Just because none of the assertions used to verify a design fail, it does not follow that the design has been fully tested. Every assertion might have passed vacuously.

One way to check that the properties are fully tested is to use a **cover** statement.

```
cover property (BillNotDispense);
```

After a simulation, the number of times the property was checked will be reported, together with the number of passes, the number of fails, and the number of vacuous passes.

We often want to know whether a *sequence* of actions has been performed.[1] For example, we might wish to check whether, after a sequence of the bill indicator

1. Sequences can also be written in separate blocks. We will not cover that aspect here.

being lit and a ticket dispensed, the vending machine lights the ready indicator. The symbol `##` is used to indicate clock cycles (according to the **clocking** block defined earlier). Thus, `##1` means one clock cycle.

```
property BillDispenseReady;
  bill ##1 dispense |=> ready;
endproperty
```

It is also possible to specify a range of cycles; thus, `##[1:3]` means "between 1 and 3 (inclusive) clock cycles." This leads to an interesting, but potentially misleading, use of sequences.

If the vending machine is waiting with its ready light lit and a $20 bill is inserted, we might expect that *eventually* a ticket will be dispensed. We do not know how many cycles this might take, because two $10 bills or one $20 bill might be inserted. It is, however, possible for a (confused) user to insert a $10 bill followed by a $20 bill. This would result in all the bills being returned. There would be nothing to stop the user from repeating this sequence ($20, $10, $20), but we might reasonably assume that eventually the user would learn! So this can be stated as a property.

```
property Eventually;
  ready && twenty |-> ##[1:$] dispense;
endproperty
```

`##[1:$]` means between one and infinity clock cycles later. This is known as a *liveness* property. In general, liveness properties should be avoided. "Eventually" can be a very long time and therefore might never fail. The property can also pass vacuously, so in practice, this property tells us nothing, while appearing to say a lot.

The assertions described thus far have been expressed in terms of input and output signals. It is also possible to use internal signals, including states. This, however, implies that either the signals must be made visible—by making them into ports—or the properties and assertions should be included in the modules themselves and not in the testbenches.

```
property StateAtoD;
  (state == A) && twenty |=> (state == D);
endproperty
```

This section has covered most of the basic syntax of assertions. To conclude, we return to the Booth multiplier example. Two properties and their associated assertions are given, that the ready signal appears at the correct time and that the correct answer is given. This last property clearly illustrates one of the great advantages of using assertions. The product generated by the multiplier can be checked by writing a simple arithmetic statement. This example also shows how an action can be associated with an assertion so that in the event of its failing, a clear message is printed. Notice that the assertions are included in a **module**, not a **program**. The point at which assertions are evaluated is described in Chapter 9.

```
module verify #(parameter AL = 8, BL = 8, QL = AL+BL)
  (input logic signed [QL-1:0] qout,
   input logic ready,
   input logic signed [AL-1:0]ain,
   input logic signed [BL-1:0] bin,
   input logic clk, load);

default clocking clock_block
 @(posedge clk);
endclocking

property LoadReady;
  load |=> ##AL ready;
endproperty

property Product;
  load |=> ##AL (qout == ain*bin);
endproperty

assert property (LoadReady);

assert property (Product)
else $error("%d * %d gives %d", ain, bin, qout);

endmodule
```

As an alternative to simulation, model checking tools can verify code using static analysis. Assertions play a key part because RTL code, for example, is checked against assertions. These can be the same as those used for simulation-based verification. Although model checking has yet to achieve wide usage, the high cost of simulation means that it is likely to play an ever-increasing role. Assertions will therefore become increasingly important in the design cycle.

Summary

Testbench writing is as important as modeling hardware. The entire SystemVerilog language can be used to write testbenches. All but the simplest testbenches should be structured to allow reuse. Stimuli, other than the clock and reset, should be included in **program** blocks to avoid synchronization problems. Constrained random test generation can give a wider variety of stimuli than simple deterministic patterns. Assertions may be used to verify responses.

Further Reading

Testbench structures are described in detail in the books by Bergeron [4] and Spear [21]. Vijayaraghavan and Ramanathan [24] provide a comprehensive description of assertion-based verification in SystemVerilog. The SystemVerilog Standard [2] gives the definitive syntax.

Exercises

8.1 Write a testbench for the sequencer from the microprocessor example of Chapter 7 that generates random opcodes and operands, such that the program is limited to the bottom half of the address space and data to the top half.

8.2 What are the benefits of using assertion-based verification?

8.3 The following piece of code is a SystemVerilog model of a counter. Write a SystemVerilog testbench for this model. Include an assertion that the ready signal is always 0 or 1.

```
module counter #(parameter N = 12)
     (input logic clk, n_reset, load, output logic ready);

int count;

always_ff @(posedge clk, negedge n_reset)
  if (!n_reset)
    begin
    count<=0;
    ready <= '1;
    end
  else
    if (load)
      count <= N;
    else if (count > 0)
      count <= count - 1;
    if (count == 0)
      ready <= '1;
    else
      ready <= '0;
endmodule
```

8.4 The following SystemVerilog assertion is intended to determine whether the counter of Exercise 8.3 is working correctly.

```
assert property (@(posedge clk) load |=> ##[1:$] ready);
```

Explain what this assertion tests and why it might be considered a poorly designed assertion. Write an assertion that checks whether the ready signal becomes true in the correct clock cycle. Write a second assertion to check that the load signal is only true when the ready signal is true.

9 SystemVerilog Simulation

SystemVerilog is a language for describing digital systems. To verify that a model is correct, a simulator may be used to animate the model. It is also important to remember that RTL synthesis attempts to generate low-level hardware that behaves in the same way as the original code. In other words, the interpretation of SystemVerilog structures for synthesis is based on the simulation model. In the first section of this chapter, the principles of digital simulation are described. The specifics of SystemVerilog simulation and techniques to improve simulation efficiency are then discussed.

9.1 Event-Driven Simulation

SystemVerilog is a language for describing digital systems; therefore, it should be of no surprise that standard event-driven logic simulation algorithms are used. Such algorithms are most easily described in terms of the simulation of structural models. Behavioral models are evaluated in much the same way, where a procedural block can be thought of as an element.

The objective of event-driven simulation is to minimize the work done by the simulator. Therefore, the state of the circuit is only evaluated when a change occurs in the circuit. It is possible to predict when the output of an element might change because we know that such a change can only occur after an input changes. If we only monitor the inputs to elements, we can only know that an output might change; the logic function of the element determines whether a change actually occurs. As we also know the delays through the element, we know when the output *might* change. Thus, an element only needs to be evaluated when it is known that its output might change but not otherwise. Nevertheless, even by predicting a possible change, it is only necessary to re-evaluate elements when the possible changes occur. By following the possible *events* through the circuit, we can minimize the computation done by the simulator. Only elements that change need to be evaluated; anything that is not changing is ignored.

The delays through elements are defined in terms of integer times. The units of time might be nanoseconds or picoseconds. As the time is incremented in discrete intervals, it is likely that, for any reasonably large circuit, more than one element will be evaluated at any one time. Equally, there may be times at which no elements are due for evaluation. This implies a form of time step control. As each element is evaluated, any change in its output will cause inputs to further elements to change, and hence the outputs of those elements may subsequently change. Clearly, it is necessary to maintain a list of which signals change and when. An *event* is therefore a new value of a signal, together with the time at which the signal changes. If the event list is ordered in time, it should be easy to add new events at the correct place in the future.

A suitable data structure for the event list is shown in Figure 9.1. When an event is predicted, it is added to the list of events at the predicted time. When an event is processed, it is removed from the list. When all the events at a particular time have been processed, that time can be removed.

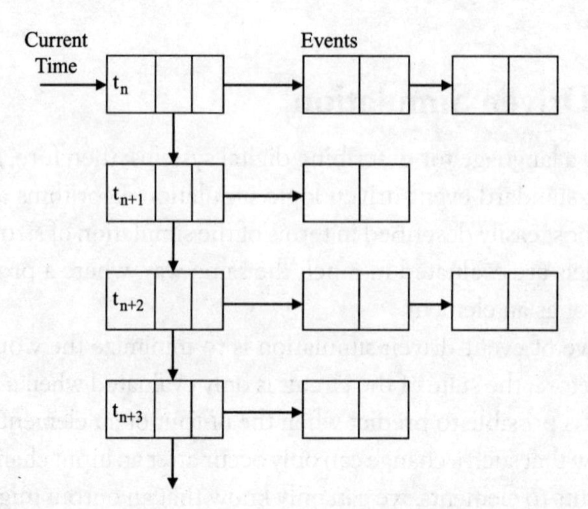

Figure 9.1 Event list.

This list manipulation is relatively easy to do in a block-structured programming language, such as C, although adding new times to the middle of a list can be awkward.

An element can be scheduled for processing when its inputs are known to change. For example, consider an AND gate with a 4-ns delay. When the signal at one input changes, it can be predicted whether the signal at the output changes depending on the state of the other inputs of the gate. If the output does change, the output event can be scheduled 4 ns later. The algorithm can be written in pseudo-C, as shown in Listing 9.1.

```
for (i = each event at current_time)
  {
  update_node(i);
  for (j = each gate on the fanout list of i)
    {
    update input values of j;
    evaluate(j);
    if (new_value(j) last_scheduled_value(j) ) {
      schedule new_value(j) at
        (current_time + delay(j) );
      last_scheduled_value(j) = new_value(j);
      }
    }
  }
```

Listing 9.1 Single-pass event scheduler.

An event is only scheduled if the new value is different from the value that has previously been scheduled for that signal. If two or more events occur on input signals to an element, more than one event may be scheduled for an output signal. It is important to know that the new value is not merely different from the present value but also from a value that might already be scheduled to be set in the future. This algorithm therefore has a disadvantage as it stands because an element is evaluated whenever an event occurs at an input. It is quite possible that two events might be scheduled for the same gate at the same time. This could lead to a zero-width spike being scheduled one delay later. Even worse, if the delays for rising and falling output differ, the presence or absence of an output pulse would depend on the order in which the input events were processed.

If zero-width pulses are to be suppressed, they can be considered as a special case of the inertial delay model, introduced in Section 3.7. Real gates require pulses of a minimum width to switch. The width of a pulse is (roughly) proportional to its energy. A real system needs a certain amount of energy to change state. A pulse with a width less than the delay is ignored. The previous algorithm can be extended to include pulse cancellation if a pulse is less than the permitted minimum width, as shown in Listing 9.2. This model assumes two-state logic. If an event is predicted at a time less than the inertial delay after the previous event for that node, this new event is not set, and the previous event is also removed. If more than two-state logic is used, the meaning of an inertial delay and hence of a canceled event must be thought about carefully. In order to cancel an event, it is necessary to search through the event lists. Event cancellation is therefore best avoided, if possible.

```
for (i = each event at current_time)
  {
  update_node(i);
  for (j = each gate on the fanout list of i)
    {
    update input values of j;
    evaluate(j);
    if (new_value(j) last_scheduled_value(j) )
     {
     schedule_time = current_time + delay(j);
     if (schedule_time last_scheduled_time(j) +
         inertial_delay(j))
       {
       cancel_event at last_scheduled_time(j);
       }
     else {
       schedule new_value(j) at schedule_time;
       }
     last_scheduled_value(j) = new_value(j);
     last_scheduled_time(j) = schedule_time;
     }
   }
 }
```

Listing 9.2 Single-pass event scheduler with inertial delay cancellation.

One further problem exists with the selective trace algorithm. A gate with a zero delay would cause an event to be scheduled at the current simulation time if an input changes. Thus, while events are being processed at the current time, new events are being added to the end of the event list. There is clearly the potential here for an infinite loop to be generated, where the simulation never advances beyond the current simulation time. In practice, the only way to avoid this problem is to count the number of events at a time point, and if they exceed some arbitrary limit, then terminate the simulation. We have already noted that the presence or absence of zero-width pulses can be dependent upon the order of evaluation of events at a time point. Consider the circuit of Figure 9.2. If both gates have a zero delay and the input changes from 0 to 1 as shown, a zero-width pulse may be generated.

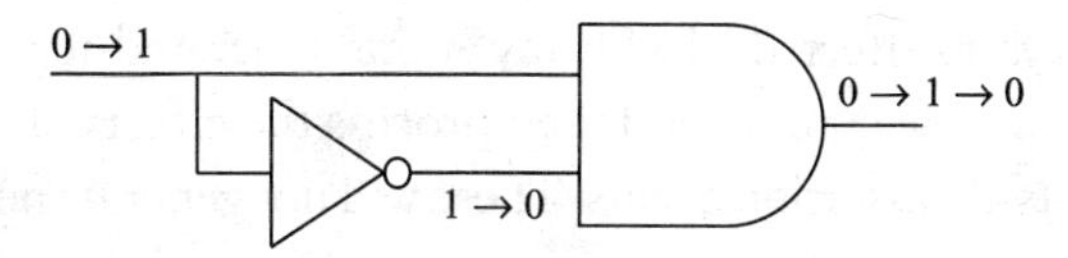

Figure 9.2 Circuit with zero delays.

If the AND gate is evaluated first, both inputs will appear to be at logic 1, so the output will become 1. The inverter is evaluated next, causing the other AND input to change. The AND gate is evaluated again and the output changes back to 0. On the other hand, if the inverter is evaluated first, both inputs to the AND gate will appear to change simultaneously when it is evaluated, and no pulse is generated. Although it is obvious here that the inverter should be evaluated first, this is not always the case, and we must assume that the order of evaluation of gates is effectively arbitrary. This arbitrariness can cause significant problems in SystemVerilog modeling.

9.2 SystemVerilog Simulation

The SystemVerilog simulation model is based upon the selective trace algorithm. The standard [2] describes a *stratified event queue* in which the event list is divided into a number of *regions*. These regions include:

- Preponed—sample stable values for later checking
- Active—blocking assignments
- Inactive—zero delay assignments
- NBA—Nonblocking assignments
- Observe—Evaluate assertions
- Reactive—Execute programs in testbenches
- Postponed—**`$strobe`** and **`$monitor`** print routines

Working backward, postponed events are created by **`$monitor`** and **`$strobe`** tasks. These cannot create new events, so they will always be executed last at a simulation time.

Reactive events refer to events generated in a **`program`**. As discussed in Chapter 8, putting stimuli into a **`program`** ensures that all stimuli are created in the same phase of the simulation cycle and not interleaved with other events. It might seem slightly strange to have such events at the end of the cycle, but as will be seen, these events can create new active events at the current simulation time.

So this ordering says, in effect, deal with any events generated at previous simulation times, then generate new stimuli, and then process the effects of these stimuli.

Observe events are assertion events. These will not generate new events. Putting these before reactive events gives us two opportunities to observe behavior—before and after stimuli are generated.

Nonblocking assign update events are created by NBAs (<=). The evaluation of the right-hand side of all NBAs is *always* done before *any* nonblocking assign updates are done. This is important as it allows sequential systems to be modeled correctly.

Inactive events are those events that are due to occur at the current time but that have been explicitly delayed. In practice, this can be done with a zero delay (#0). As a general guideline, do not use zero delays! A zero delay does not represent real hardware (nor a useful testbench construct). Therefore, you are simply trying to fool the simulator. Unless you know exactly what you are doing, it will probably fool you!

Active events are, at first glance, just blocking assignments. As will be seen in a moment, however, as the list of blocking assignment events is exhausted, subsequent event lists become active. So, for example, processing a list of nonblocking assignment events could produce new active events (from blocking assignments, for example), which are put directly back into the active event queue.

Finally, preponed events are sampling events.

The relationship between the regions is illustrated in Figure 9.3.

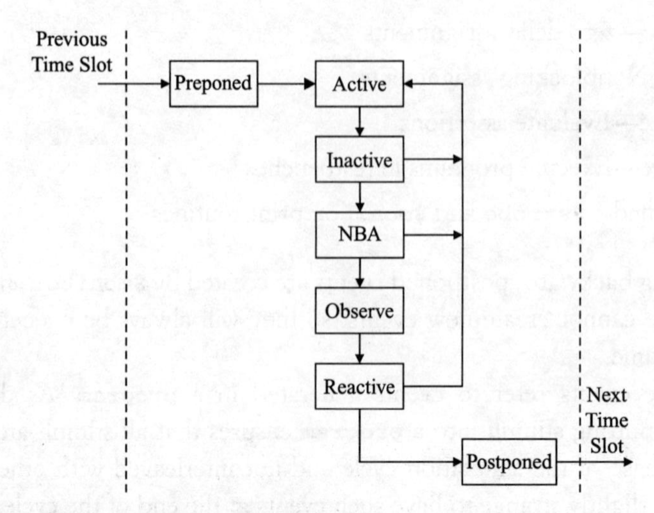

Figure 9.3 SystemVerilog stratified event queue.

The pseudo-code of Listing 9.3, adapted from the SystemVerilog standard, describes the simulation cycle. Each iteration of the loop is one cycle. T is the current simulation time.

```
while (there are events) {
  if (no active events) {
    if (there are events in the next list) {
      activate all events in the next list;
    else { /*if no more events at the current time*/
      advance T to the next event time;
      }
    }
    E = any active event;
    if (E is an update event) {
      update the modified object;
      add evaluation events for sensitive processes
        to event queue;
      }
    else { /* shall be an evaluation event */
      evaluate the process;
      add update events to the event queue;
      }
  }
```

Listing 9.3 SystemVerilog simulation cycle.

Only active events are processed, but a new event may be assigned to one of regions listed previously.

From the pseudo-code and this description, it can be seen that the list of active events is one of the lists that has been created during some previous simulation cycle, together with any (active) events that are generated during the current cycle.

Events may be processed from the active event list in any order (or to think of it another way, events can be added to the event lists in any order). This is the fundamental cause of non-determinism in SystemVerilog.[1] We can be sure of only two things:

1. Statements between **begin** and **end** will be executed in the order stated.
2. Nonblocking assignments will be performed after blocking assignments.

1. VHDL experts may be looking for *delta delays*. There is no such thing in SystemVerilog. The existence of the delta delay means that a VHDL simulation is deterministic and repeatable between simulators. The absence of delta delays in SystemVerilog means that simulations may not be deterministic and may not be repeatable between different simulators.

Everything else is indeterminate. Moreover, the execution of a procedural block can be interrupted to allow another procedural block to be executed. The skill in writing SystemVerilog code is therefore to ensure that this non-determinism does not matter. If the code is badly written, a *race* condition is likely to result—that is, a situation where the procedure writing a value and the procedure reading that value are racing each other. Depending upon which completes first, either the original or the updated value may be read.

9.3 Races

Perhaps the best way to understand races is by examples. The first example comes from the standard [2].

```
assign p = q;

initial
  begin
  q = 1;
  #1 q = 0;
  $display(p);
  end
```

Because the execution of the initial procedure can be interleaved with the assign statement, the value of p that is displayed could be 1 or 0. Either is "correct," and different simulators may give different results.

The second example was shown in Chapter 5. This is an example of an assignment to a variable from two different procedures.

```
initial clock = '0;
always #10ps clock = ~ clock;
```

The assumption (presumably) is that the initial procedure is evaluated first and then the clock changes every 10 ps. If the **always** procedure is evaluated first, the assignment is redundant because it is superseded by the assignment in the initial procedure. In either case, a clock waveform is generated, but the two cases are out of phase.[2]

This ambiguity can be overcome by changing the assignment in the always procedure to an NBA. By definition, the NBA is evaluated after the blocking assignment

2. It is quite easy to demonstrate this problem by reversing the order of the statements.

in the initial procedure. This is, however, bad practice because one variable is assigned values from two procedures.

A third example of a race is shown next.

```
always @(posedge clock)
  b = c;

always @(posedge clock)
  a = b;
```

This is supposed to model two flip-flops connected in series. If the procedures are evaluated in the order written, at a rising clock edge, the value of c will be copied to b. A new event will be scheduled at the current time, causing that same value to be copied to a. This is clearly not the intended behavior. If the procedures are evaluated in the opposite order, the correct behavior is modeled.

9.3.1 Avoiding Races

In order to achieve deterministic behavior, there are several rules that should be followed when writing models and testbenches.

Do not assign to the same variable from two or more procedures. Not only is contention liable, but as can be seen from the second example, multiple assignments can cause ambiguous behavior. Part of the problem is the keyword **initial**. Procedures defined with the **initial** keyword are executed once; they are not *initialization* blocks.

Use NBAs for modeling sequential logic. The third example evaluates correctly if the two assignments are made nonblocking, irrespective of the order of evaluation. This is because NBAs are evaluated last and cannot influence each other.

Conversely, use blocking assignments to evaluate combinational logic. Assignments made in combinational logic models are supposed to take immediate effect. The use of NBAs would often be confusing (and wrong).

Some experts argue that blocking and nonblocking assignments should not be mixed in the same procedure. Certainly, it is not a good idea to use both types of assignment to the same variable as some synthesis tools will object (although such constructs are syntactically valid). This restriction has generally been followed in these notes, but it is possible to mix the two types of assignments such that blocking assignments are used for all combinational logic and NBAs used for all registered outputs.

Do not use zero delays (#0). They are not necessary and will cause confusion, as explained previously.

9.4 Delay Models

SystemVerilog provides five ways to model delays:

(1) Left-hand side (LHS) of blocking assignments:

```
#5 a = b;
```

(2) Right-hand side (RHS) of blocking assignments:

```
a = #5 b;
```

(3) LHS of NBAs:

```
#5 a <= b;
```

(4) RHS of NBAs:

```
a <= #5 b;
```

(5) LHS of continuous assignments:

```
assign #5 a = b;
```

None of these constructs is *needed* for RTL modeling. (1) and possibly (3) and (4) are useful for testbench writing. (4) is a transport (pure) delay that can be used to model delays in sequential logic (such as clock to output delay in a flip-flop). (5) is an inertial delay that can be used to model delays in combinational logic. To understand why some forms are useful and others are not, we need to understand what precisely occurs in each case.

In form (1), the simulator waits for, for example, five time units and then executes the assignment. Any changes to inputs during this wait period are ignored. Clearly, this does not model real hardware, but is useful for describing a waveform in a testbench.

Form (2) causes the present value of the RHS to be scheduled for assignment at some point in the future. This is a transport or pure delay—every change, no matter how rapid, is transmitted to the output. This could be used to model, say, a transmission line. This is not particularly useful for testbench design.

Form (3) again causes a five-unit wait before assigning the then present value of the RHS. Again, any intermediate changes are ignored and so this does not model real hardware. This can be used in testbenches, but has no advantage over form (1).

As noted, (4) can be used to model delays in sequential logic. A transport delay is appropriate here, in contrast to the last form.

The final form, (5), models an inertial delay. Therefore, this form of delay best models combinational logic.

9.5 Simulator Tools

The major EDA tool vendors each have their own SystemVerilog simulators. Additionally, there are a number of open source simulators available for older versions of Verilog. Although these tools have been written by different people at different times, they all implement versions of the simulation algorithm described earlier in this chapter. Therefore, in general terms, the simulators behave in broadly the same ways. There is indeterminacy in the simulation algorithm and therefore care has to be taken to ensure that models and testbenches are written in such a way that the indeterminacy does not matter.

All simulators also go through a similar sequence of operations before simulation. SystemVerilog models have to be compiled and elaborated. *Compilation* is the process of translating SystemVerilog code from text into a binary form that can be executed by the simulator. This is closely analogous to the task performed by the compiler of a software language, such as C. Because SystemVerilog is weakly typed, it is possible to write ambiguous and poorly defined code. To some extent, this can be alleviated by using a *linting* tool to analyze files before compilation to identify badly written code.

The next step, *elaboration*, is similar to linking in a software environment. Typically, two or more SystemVerilog modules in a number of source files are compiled independently. Therefore, during elaboration, model instances have to be bound to model declarations and the model hierarchy established. Moreover, parameter values have to be resolved and nets connected together. If parts of the model description are missing or if there is ambiguity, simulation cannot go ahead, and the user needs to be told.

Following elaboration, a binary representation of the complete model hierarchy is loaded into the simulator (or linked with the simulator to form a standalone executable).

The user-interface of the simulator acts as a SystemVerilog process. The user can start and stop the simulation and inspect signals. The most immediately obvious aspect of a simulator is the waveform display tool, which allows inspection of the function of the model being simulated. For example, the waveforms generated by the testbench for the N-bit adder, given in Section 4.7, are shown in Figure 9.4.[3]

The other main tool included in most simulators is the interactive debugger. This can be used in exactly the same way as software debuggers. Breakpoints can be

3. These waveforms were displayed using an open source waveform tool—GTKWave.

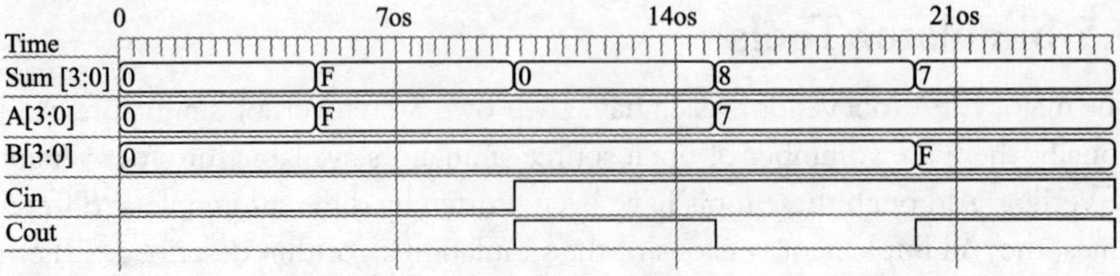

Figure 9.4 Waveforms generated by simulation of an N-bit adder.

inserted in the source code to suspend the simulation at a particular line. The values of variables and nets can be monitored. This can be coupled with a coverage tool to determine exactly how many times (if at all) a particular line is executed. All of these tools differ in the exact implementation and interface and therefore details will not be given here. It is, however, worth investing the time in finding out how to exploit the power of such tools because that time will be paid off in much shorter development and debugging times.

Summary

The SystemVerilog simulation model has five distinct event queues. The ordering of events within these queues is not defined. This leads to non-determinism in SystemVerilog simulations. It is possible to make SystemVerilog simulations deterministic by adopting well-tried design styles. These styles are also appropriate for RTL synthesis. There are five possible styles of delay modeling, but only two of these are useful for RTL modeling and one other is best suited to testbench writing.

Further Reading

Logic simulation is described in the books by Miczo [15] and Abramovici, Breuer, and Friedman [3]. The SystemVerilog simulation cycle is described in the SystemVerilog standard [2].

Exercises

9.1 Explain the SystemVerilog stratified event queue.

9.2 The following piece of code is a SystemVerilog model of two flip-flops. Explain why the behavior of this model might not be the same in two different simulators.

```
always_ff @(posedge clock)
  a = b;

always_ff @(posedge clock)
  b = c;
```

9.3 Explain what is meant by "inertial" and "transport" delays. Give an example of how each would be described in a SystemVerilog model.

9.4 A SystemVerilog model of two interconnected state machines is shown in the following. What is the likely purpose of the two zero delays (#0) in the model? Rewrite this model such that the non-deterministic features are removed.

```
always_ff @(posedge clock,
            negedge n_reset)
  if (!n_reset)
    begin
    enable = '0;
    state = s0;
    end
  else
    case (state)
      s0: begin
            enable = '0;
            if (start)
              state = s1;
            end
      s1: begin
            #0 enable = '1;
            if (timed)
              state = s0;
            end
    endcase

always_ff @(posedge clock,
            negedge n_reset)
  if (!n_reset)
    begin
    count = 0;
    timed = '0;
    end
  else
    begin
    if (enable)
```

```
    begin
    count++;
    timed = '0;
    end
  if (count == 255)
    begin
    #0 timed = '1;
    count = 0;
    end
  end
```

10 SystemVerilog Synthesis

Verilog was originally designed as a hardware *description* language. In other words, the language was designed to model the behavior of existing hardware, not to specify the functionality of proposed hardware. Moreover, when the Verilog language was originally designed, there were no automatic synthesis tools in widespread use. Therefore, the meaning of different SystemVerilog constructs in hardware terms was derived some years after the language was standardized. The consequence of this is that parts of SystemVerilog are not suitable for synthesis.

At this point, we should define what we mean by the term *synthesis*. The long-standing objective of design automation tool development has been to compile high-level descriptions into hardware in much the same way that a computer software program is compiled into machine code.

Figure 10.1 shows a simplified view of the design process. After a specification has been agreed upon, a design can be partitioned into functional units (architectural design). Each of these functional units is then designed as a synchronous system. The design of these parts can be done by hand, as described in Chapter 6. Thus, a state machine is designed by formulating an ASM chart, deriving next state and output equations, and implementing these in combinational logic. At this point, the gates and registers of the design can be laid out and wired up on an integrated circuit or PLD.

Figure 10.1 shows how synthesis tools can automate parts of this process. RTL synthesis tools take a SystemVerilog description of a design in terms of registers,

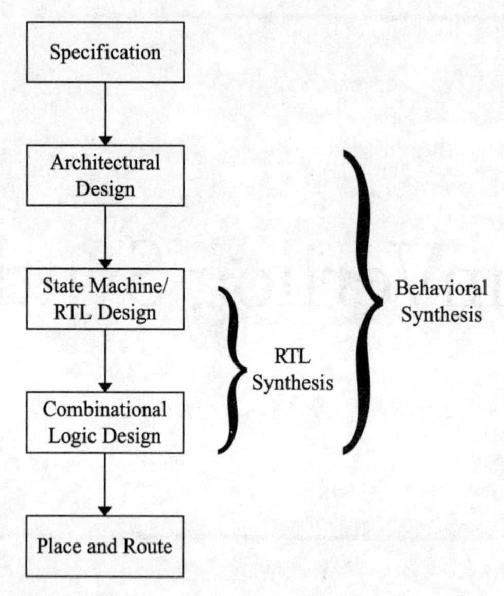

Figure 10.1 High-level design flow.

state machines, and combinational logic functions and generate a netlist of gates and library cells. As we will see, the SystemVerilog models described in Chapters 4, 5, 6, and 7 are mostly suitable for RTL synthesis. Behavioral synthesis tools, on the other hand, take algorithmic SystemVerilog models and transform them into gates and cells. The user of a behavioral synthesis system would not have to specify clock inputs, for instance, simply that a particular operation has to be completed within a certain time interval. RTL synthesis tools are gaining widespread acceptance; behavioral synthesis tools are still relatively rare. Although this chapter is about RTL synthesis, it is likely that in a few years behavioral synthesis tools will be widely accepted, in a manner analogous to the way that high-level software programming languages such as Java are coming to replace lower-level languages such as C.

The last stage of the synthesis process, place and route, is carried out by separate specialized tools. In the case of programmable logic, the manufacturers of the PLDs often supply these tools.

10.1 RTL Synthesis

The functions carried out by an RTL synthesis tool are essentially the same as those described in Chapter 6. The starting point of the synthesis process is a model (in SystemVerilog) of the system we wish to build, described in terms of combinational and sequential building blocks and state machines. Thus, we have to know all the inputs and outputs of the system, including the clock and resets. We also have to know the number of states in state machines—in general, RTL synthesis tools do not perform state minimization. From this we can write SystemVerilog models of the parts of our system. In addition, we may wish to define various *constraints*. For instance, we might prefer that a state machine be implemented using a particular form of state encoding. We almost certainly have physical constraints such as the maximum chip size and hence the maximum number of gates in the circuit and the minimum clock frequency at which the system should operate. These constraints are not part of SystemVerilog, in the sense that they do not form part of the simulation model, and are often unique to particular tools, but *may* be included in the SystemVerilog description.

The IEEE standard 1364.1-2002 defines a subset of Verilog for RTL synthesis. The purpose of this standard is to define the minimum subset that can be accepted by *any* synthesis tool. SystemVerilog is based on the 2001 enhancements to Verilog. The synthesizable examples in this book are designed to conform to the 1364.1 standard.

10.1.1 Non-Synthesizable SystemVerilog

In principle, most features of SystemVerilog could be translated into hardware. In general, those parts of SystemVerilog that are not synthesizable are constructs in which exact timing is specified and structures whose size is not completely defined. Poorly written SystemVerilog may result in the synthesis of unexpected hardware structures. These will be described later.

The following SystemVerilog constructs are either ignored or rejected by RTL synthesis tools.

- All delay clauses (e.g., #10). Delays are *simulation* models. A model can be synthesized to meet various *constraints*, but cannot be synthesized to meet some exact timing model. For instance, it is not possible to specify that a gate will have a delay of exactly 5 ns. It is reasonable, on the other hand, to require a synthesis tool to generate a block of combinational logic such that its total delay is less than, say, 20 ns.
- File operations suggest the existence of an operating system. Hence, file operations cannot be synthesized and would be rejected by a synthesis tool.
- Real data types are not inherently *unsynthesizable*, but will be rejected by synthesis tools because they require at least 32 bits, and the hardware required for many operations is too large for most ASICs or FPGAs.
- Initial blocks will be ignored. Hardware can't exist for a limited period of time and then disappear!

10.1.2 Inferred Flip-Flops and Latches

It is important to appreciate that synthesis tools (like most computer software) are basically stupid. While there are reserved words in SystemVerilog to specify whether a model is combinational or sequential, inconsistent models may only generate warnings. Therefore, the fundamental problem with synthesizing SystemVerilog models is to ensure that the hardware produced by the synthesis system is what you really want. One of the most likely "errors" is the creation of additional flip-flops or latches. Therefore, in this section, we will describe how the existence of flip-flops and latches is inferred.

A flip-flop or latch is synthesized if a net or register holds its value over a period of time. In SystemVerilog, a net holds its value until it is given a new value. A flip-flop or latch is created implicitly if some paths through a procedure have assignments to a net or register while others do not. This typically happens if a **case** statement or an **if** statement is incomplete in the sense that one or more branches does not contain an assignment to a register while other branches do contain such an assignment, or if the **if** statement does not contain a final **else**.

The term "flip-flop" refers here to a memory element triggered by an edge of the clock. "Latch" refers to a level-sensitive device, controlled by some signal other than the clock. Thus, a flip-flop would be created if the event list of a block has a **posedge** or **negedge** expression, while a latch would be created if the level value of a net were used instead.

In principle, therefore, procedural blocks with various edge-triggered and level-sensitive expressions could be synthesized. In practice, synthesis tools recognize a small number of fairly simple patterns, as shown in the rest of this section. These examples can act as templates for larger examples. It should be noted that in all these examples, the net names are not significant to the synthesis tool. Thus, a clock net might be called "Clock" or "Clk1" or, with equal validity, "Data." Note, however, that good software engineering practice should be applied, and *meaningful* identifiers should be used for the benefit of your readers.

10.1.2.1 Level-Sensitive Latch If we really want to create a latch, we can specify it by using the particular form of the always block:

```
always_latch
  if (Ctrl)
    Z <= A;
```

A general always block can also be interpreted as a latch. The following example shows the SystemVerilog that would be interpreted to specify a level-sensitive latch by an RTL synthesis tool.

```
always @(Ctrl or A)
  if (Ctrl)
    Z <= A;
```

The **always** statement has an event list containing the net (or register) `Ctrl` and the net, `A`, which is assigned to the output. Therefore, the statement is executed when `Ctrl` or `A` changes. `Z` is assigned the value of `A` if `Ctrl` has just changed to a 1. While `Ctrl` is 1, any change in `A` is transmitted to the output. Otherwise, no assignment to `Z` is specified. Therefore, it may be inferred that `Z` holds its value, and hence it is inferred that `Z` is a registered net. This inference can be avoided if the **else** clause is included:

```
always @(Ctrl or A)
  if (Ctrl)
    Z <= A;
  else
    Z <= 1'b0;
```

The value of `Z` is therefore `Ctrl AND A`. On the other hand, specifying a block as a latch when it is not should generate a warning.

```
always_latch
  if (Ctrl)
    Z <= A;
  else
    Z <= 1'b0; // This is inconsistent
```

`Case` statements are interpreted in a similar manner.

```
always @(Sel, A, B)
  case (Sel)
    2'b00 : Y <= A;
    2'b10 : Y <= B;
    default;
  endcase;
```

The **default** clause covers the patterns 01 and 11 (and combinations with X and Z, although they are irrelevant to synthesis). If the default clause were omitted, the **case** statement would still be syntactically correct. When `Sel` is one of these

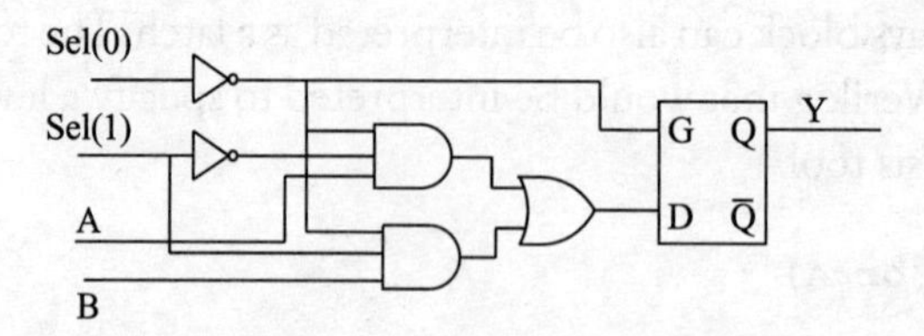

Figure 10.2 Circuit synthesized from an incomplete **case** statement.

two missing patterns, Y is assumed to hold its value. Hence, the circuit of Figure 10.2 is synthesized.

Note that the latch used in these examples would be taken from a library. Such elements cannot be synthesized from first principles by a synthesis tool. The continuous assignment statement

```
assign y = E ? D : y;
```

in which a signal appears on both the left- and right-hand sides of the net assignment, might be synthesized to the circuit of Figure 10.3. This is apparently functionally correct, but it contains a potential hazard and is therefore a poor latch design. The synthesis standard disallows such constructs.

10.1.2.2 Edge-Sensitive Flip-Flop As described in Chapter 5, edge-sensitive behavior may be modeled by putting a **posedge** or **negedge** expression in an event list:

```
always_ff @(posedge clk)
  q <= d;
```

or

```
always @(posedge clk)
  q <= d;
```

The **posedge** and **negedge** statements are interpreted by a synthesis system to model edge-sensitive behavior. Hence, net assignments that can only be reached by fulfilling an edge-sensitive condition will be interpreted as assignments to registered

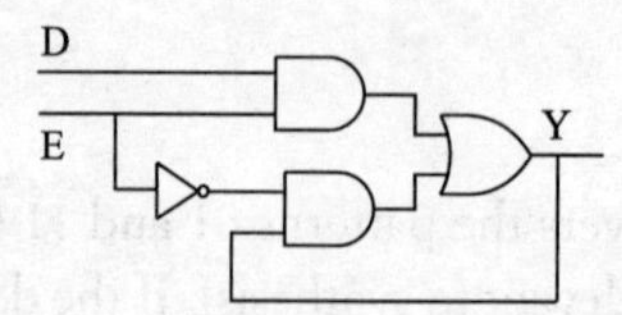

Figure 10.3 Asynchronous circuit synthesized from a feedback assignment.

nets. It should be remembered that the net name itself is not meaningful to the synthesis tool.

Asynchronous sets and resets are modeled by including the active edge in the event list:

```
always_ff @(posedge clk, posedge reset)
  if (reset)
    q <= 1'b0;
  else
    q <= d;
```

This structure would be interpreted, correctly, as a positive-edge triggered flip-flop with an active high asynchronous reset. The reset is tested before the clock and therefore has an effect irrespective of the clock. The clock net to which the flip-flop is edge-sensitive should be tested in the last branch of the **if** statement. Similarly, synchronous sets and resets and clock enable inputs as described in Chapter 5 will be correctly interpreted by an RTL synthesis tool.

We saw in Chapter 9 that the SystemVerilog simulation model means that non-blocking assignments do not take effect until all other events have been processed at the current simulation time. Blocking assignments, without delays, on the other hand take immediate effect. The synthesized forms of nonblocking and blocking assignments should therefore be different. The following fragment of SystemVerilog synthesizes to the structure shown in Figure 10.4.

```
always_ff @(posedge clock)
  begin
  P <= A & B;
  Z <= P | C;
  end
```

In the first NBA, `P` is given a value. When `P` is referenced in the second assignment, the new value of `P` has not yet taken effect. Therefore, the previous value of `P` is used. The value of `P` (and of `Z`) is not updated until the procedure resumes, at the next clock edge. Therefore, `P` behaves exactly as if its value were stored in a flip-flop.

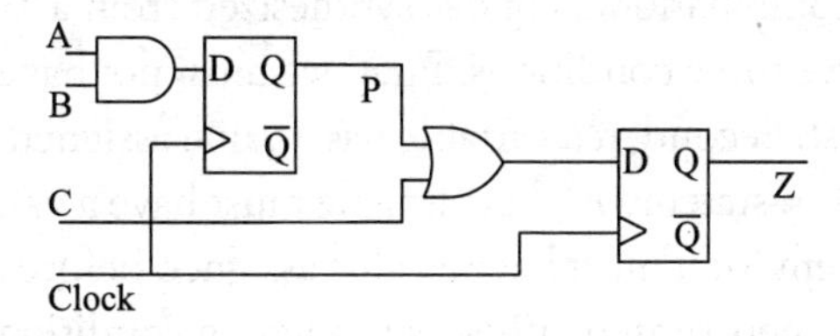

Figure 10.4 Circuit synthesized by NBAs.

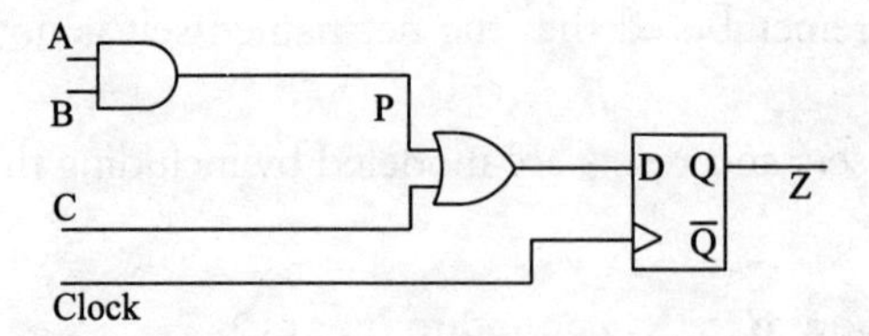

Figure 10.5 Circuit synthesized using a blocking assignment.

By contrast, a blocking assignment takes effect immediately. Therefore, the following piece of code, in which `P` is assigned a value through a blocking assignment, is synthesized to the structure of Figure 10.5.

```
always_ff @(posedge clock)
  begin
  P = A & B;
  Z <= P | C;
  end
```

In general, use NBAs to model edge-triggered flip-flops. You can use blocking assignments to model temporary variables, such as P in the last example, but SystemVerilog does not allow these temporary variables to be distinguished from other registers. Therefore, to avoid ambiguity and potential race problems, do not mix blocking and nonblocking assignments in the same procedural block. Note also that according to IEEE standard 1364.1, the event list should only contain edge-sensitive events.

10.1.3 Combinational Logic

In general, if a piece of hardware is not a level-sensitive or edge-sensitive sequential unit, it must be a combinational unit. Therefore, a SystemVerilog description that does not fulfill the conditions for synthesis to level-sensitive or edge-sensitive sequential elements must by default synthesize to combinational elements. Hence, the problem of describing combinational hardware in SystemVerilog is to ensure that we do not accidentally cause the synthesis tool to infer the existence of registers.

To ensure that combinational logic is synthesized from a SystemVerilog procedure, we must observe three conditions. First, we must not have any edge-triggered events in the event list. Second, if a variable has a value assigned in one branch of an **`if`** statement or a **`case`** statement, that variable must have a value assigned in every branch of the statement (or it must have a value assigned before the branching statement). Finally, all the nets sensed either as branching conditions or in assignments must be included in the event list of the process, if used.

For example, the following is a model of a state machine with two states, two inputs, and two outputs.

```
module Fsm (output logic OutA, OutB,
            input Clock, Reset, InA, InB);

  enum {S0, S1, S2} PresentState;

  always_ff @(posedge Clock or posedge Reset)
    if (Reset)
      PresentState <= S0;
    else
      case (PresentState)
        S0: begin
              OutA <= 1'b1;
              if (InA)
                PresentState <= S1;
              end
        S1: begin
              OutA <= InB;
              OutB <= 1'b1;
              if (InA)
                PresentState <= S2;
              end
        S2: begin
              OutB <= InA;
              PresentState <= S0;
              end
      endcase
endmodule
```

Although this is an acceptable simulation model, if it were synthesized, `OutA` and `OutB` would be registered in addition to `PresentState` because they have values assigned to them within an edge-triggered procedure. Thus, we can divide the model into two procedures, one combinational and one sequential. We will use blocking assignments in the "combinational" procedure to ensure that all the values are updated before they are read into registers.

```
module Fsm (output logic OutA, OutB,
            input Clock, Reset, InA, InB);

  enum{S0, S1, S2} PresentState, NextState;

always_ff @(posedge Clock or posedge Reset)
  if (Reset)
    PresentState <= S0;
```

```
  else
    PresentState <= NextState;

always_comb
  case (PresentState)
    S0: begin
        OutA = 1'b1;
        if (InA)
          NextState = S1;
        else
          NextState = S0;
        end
    S1: begin
        OutA = InB;
        OutB = 1'b1;
        if (InA)
          NextState = S2;
        else
          NextState = S1;
        end
    S2: begin
        OutB = InA;
        NextState = S0;
        end
  endcase
endmodule
```

This will, again, simulate as a state machine giving apparently correct behavior. When synthesized, however, `OutA` and `OutB` will be registered through asynchronous latches because in state `S0` no value is assigned to `OutB` and hence `OutB` holds onto its value. Similarly in state `S2`, no value is assigned to `OutA`. This should generate warnings.

This error can be resolved by explicitly including an assignment to both `OutA` and `OutB` in every branch of the **case** statement. Alternatively, both signals can be given default values at the start of the procedure:

```
always_comb
  begin
  OutA = 1'b0;
  OutB = 1'b0;
  case (PresentState)
    S0: begin
        OutA = 1b'1;
        if (InA)
          NextState = S1;
        else
```

```
            NextState = S0;
          end
      S1: begin
          OutA = InB;
          OutB = 1b'1;
          if (InA)
            NextState = S2;
          else
            NextState = S1;
          end
      S2: begin
          OutB = InA;
          NextState = S0;
          end
  endcase
  end
```

This procedure now synthesizes to purely combinational logic, while the other procedure synthesizes to edge-triggered sequential logic.

Note, however, that it is not essential to use the **always_comb** reserved word. It is also possible to use an **always** block, with a default event list:

```
always @(*)
```

It is also possible to list those signals that should cause the block to be evaluated. For example, suppose that the block were specified with:

```
always @(PresentState)
```

Most synthesis tools would (or should) give a warning, however. A piece of combinational logic will be synthesized with three inputs (`PresentState`, `InA`, and `InB`) and three outputs (`NextState`, `OutA`, and `OutB`). Hence, a change at any of the inputs could cause a change at an output. If the SystemVerilog model has only one signal in its event list (`PresentState`), this model and the synthesized circuit would behave differently when simulated. To avoid this, all the signals to which the combinational logic is sensitive should be included in the event list.

The style of coding will also influence the final hardware. For example, nested **if** ... **else** blocks, such as the priority encoder of Section 4.3, will tend to result in priority encoding and hence long chains of gates and large delays. On the other hand, **case** statements such as the state machine will tend to be synthesized to parallel multiplexer-type structures with smaller delays. (However, see Section 10.2.3.) Similarly, shift operations will result in structures simpler than multiplication and division operators.

Table 10.1 Summary of RTL Synthesis Rules

	Event List	Branches
Combinational logic	All inputs in event list (nets and registers on RHS of assignments and used in `if` and `case` statements) or use `always_comb`	Complete (or default values)
Latches	All inputs in event list (nets and registers on RHS of assignments and used in `if` and `case` statements) or use `always_latch`	Not complete
Flip-flops	Edge-sensitive clock, set and reset only	Not complete

10.1.4 Summary of RTL Synthesis Rules

It is easy to make mistakes and to accidentally create latches when combinational logic is intended (or worse, to deliberately create latches when you really want a flip-flop—see Section 6.5.4). Table 10.1 summarizes the rules for creating combinational and sequential logic from processes.

There is one further rule that applies to all synthesizable logic: Do not assign a value to a net or variable in two or more procedures. The only exception to this rule is the case of three-state logic, as in the bus in the microprocessor example of Chapter 7. You should be able to draw a block diagram of your design, with each procedure represented by a box. If two boxes appear to be driving the same wire, you have done something wrong. (Indeed, if you can't draw the block diagram, you have made a really serious mistake!)

10.2 Constraints

For any non-trivial digital function, there exist a number of alternative implementations. Ideally, a digital system should be infinitely fast, infinitesimally small, consume no power, and be totally testable. In reality, of course, this ideal is impossible. Therefore, the designer has to decide what his or her objectives are. These objectives are expressed to the synthesis tool as *constraints*. Typically, a design has to fit on a particular FPGA and has to operate at a particular clock frequency. Thus, two constraints of area and speed have to be specified. It is possible that these constraints will be in conflict. For example, a design may fit on a particular FPGA, but not work at the desired speed—to reach the desired speed may require more logic and hence more area, as illustrated in Figure 10.6. Assuming that CMOS logic is used and that

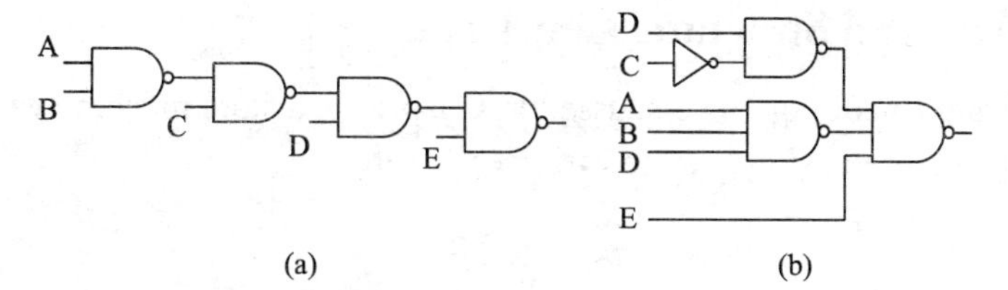

Figure 10.6 Two versions of a combinational circuit: (a) smaller, slower; and (b) larger, faster.

the gate delays are identical, the circuit of Figure 10.6(a) needs 16 transistors and a maximum delay of 4 units, while the circuit of Figure 10.6(b) requires 18 transistors and has a maximum delay of 3 units.

10.2.1 Attributes

Synthesis constraints can be expressed in two ways: as SystemVerilog attributes in the model description or as some other format in a separate file. The 1364.1 IEEE standard defines 16 attributes that can be included in the SystemVerilog description. In general, attributes are used to pass information to synthesis tools, but are ignored by simulators.

For example, 1364.1 defines one attribute for specifying the state encoding:

```
typedef enum{S0, S1, S2} fsm_state;
(* synthesis, fsm_state="onehot" *)
fsm_state present_state;
```

This might instead be expressed in a separate constraints file using a format like:

```
define_attribute fsm_state present_state "onehot"}
```

Other example attribute definitions could be as follows:

```
(* synthesis, black_box *)}

(* synthesis, implementation="ripple" *)}
```

In general, the type and format of constraints are unique to particular synthesis tools; in the following sections we discuss only the general types of constraints that can be specified.

10.2.2 Area and Structural Constraints

10.2.2.1 State Encoding As discussed in Chapter 6, a state machine with s states can be implemented using m state variables, where

$$2^{m-1} < s \leq 2^m$$

There are $\frac{(2^m)!}{(2^m-s)!}$ possible state assignments. There is no method for determining which of these assignments will result in minimal combinational next state logic. In addition, other non-minimal state encoding schemes, such as one-hot, exist. No RTL synthesis tools attempt to tackle the general state assignment problem. Heuristic methods may be able to choose either a binary counting sequence or one-hot encoding. Therefore, one design constraint that can be specified is the state encoding method, either using the IEEE 1364.1 style or by specifying the code with a keyword, as shown previously.

10.2.2.2 Resource Constraints The use of a particular technology may constrain the type of structures that can be created. Features of different FPGA technologies are discussed later in this chapter. Having selected a particular technology, a range of different-sized devices may exist, and very often it is desirable to select the smallest possible. Thus, the specification of a particular device is a constraint on the synthesis process.

As a single ASIC or FPGA has to be connected via a printed circuit board to other devices, the functionality of each pin may have to be determined in advance of the synthesis. Therefore, another constraint is the association of a signal with a particular pin.

Under some circumstances, complex logic blocks may be reused. For example, the following piece of code can be implemented with two adders or with one adder and two multiplexers.

```
if (Select)
  q = a + b;
else
  q = c + d;
```

A synthesis constraint can choose whether resources may be shared, either at a local level or globally. Such choices have implications for both the area and speed of the final design. The following attribute can be attached to a module:

```
(* synthesis, op_sharing *)
```

Finally, it may be desirable to describe a function in SystemVerilog in order to verify the correct operation of the rest of the system, but when the system is

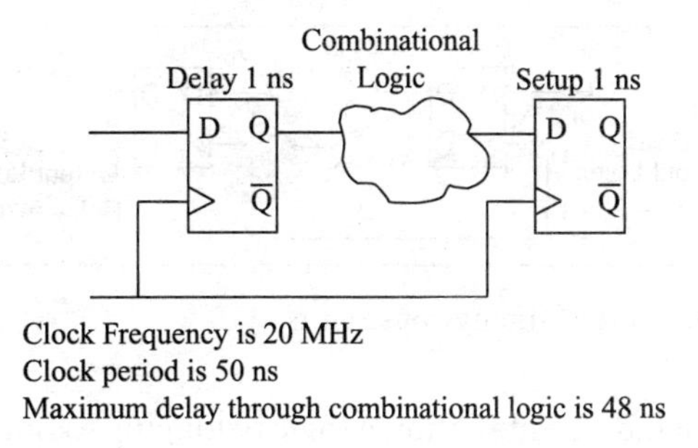

Figure 10.7 Basic timing constraint.

synthesized we would rather use a predefined library component to implement that function instead of synthesizing the function from first principles. Therefore, we can designate that a particular unit is a "black box" that we will incorporate from a library, for example,

```
(* synthesis, black_box *)
```

10.2.2.3 Timing Constraints If we want a circuit to operate synchronously with a clock at a particular frequency, say 20 MHz, we know that the maximum delay through the state registers and the next state logic is the reciprocal of the clock frequency, in this case 50 ns. Therefore, a constraint on the synthesis tool can be expressed as the clock frequency or as the maximum delay through the combinational logic, as shown in Figure 10.7.

The difficulty, from the synthesis point of view, with this approach is that the delay through the combinational logic can only be estimated. The exact delay depends on how the combinational logic is laid out, and hence the delay depends on the delay through the interconnect. Therefore, the synthesis is performed using an estimate of the likely delays. Having generated a netlist, the low-level place and route tool attempts to fit the design onto the ASIC or FPGA. The place and route tool can take into account the design constraint—the maximum allowed delay—and the delays through the logic that has been generated. At this stage, it may become apparent that the design objective cannot be achieved, so the design would have to be synthesized again with a tighter timing constraint to allow for the extra time in the routing. This can mean that the final goal is never reached. To speed up hardware more operations are performed concurrently, which means that the design is larger. Hence, the design is harder to place and route, and hence the routing delays increase, *ad infinitum*.

More specific timing, constraints can be applied to selected paths. If a design is split between two or more designers, the signal path between registers in two parts

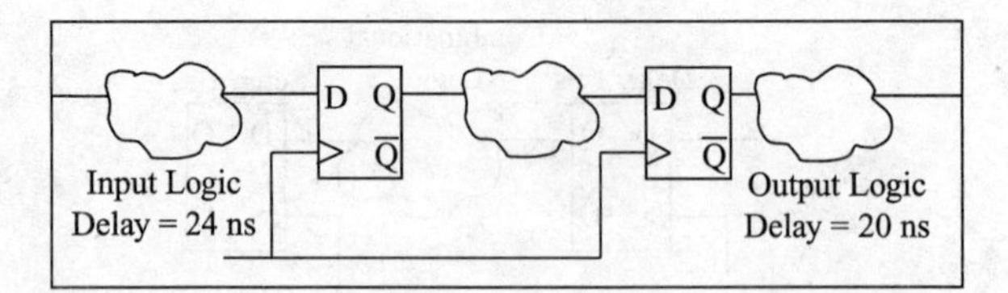

Figure 10.8 Input and output timing constraints.

of the design may include combinational logic belonging to both parts of the design. If both parts of combinational logic were each synthesized without allowing for the existence of the other, the total delay between registers could be greater than one clock period. Therefore, timing constraints can be placed upon paths through the input and output combinational logic in a design, as shown in Figure 10.8.

10.2.3 `full_case` and `parallel_case` Attributes

Many SystemVerilog designers attach the `full_case` and `parallel_case` attributes to case statements, without thinking. In general, this is a bad practice and these directives *should not* be used. The attributes only apply to synthesis and may cause the synthesized hardware to have different functionality from that simulated in RTL. To understand why these attributes may cause problems, we need to understand what the terms "full case" and "parallel case" mean.

In SystemVerilog, each case item (the term to the left of the colon in each branch) is tested in turn against the case expression. A full case statement is one in which every combination of 0, 1, z, x in the case expression can be matched against one (or more) case item(s). This applies to casez (and casex) statements, in which there are don't care terms. It is not required that a case statement is full, but, on the other hand, this condition can be achieved simply by including a default item in the case statement. If a case statement is not full, and the uncovered alternatives include combinations of 0 or 1, the correct interpretation in synthesis would be to create a latch. For example, the following is a simplified version of the priority encoder from Chapter 5. If the default item were omitted, the pattern 4b'0000 (or, indeed, 4b'000z or any pattern that included an x) would not be matched and the case statement would not be full.

```
always @(a)
  casez (a)
    4'b1??? : y = 2'b11;
    4'b01?? : y = 2'b10;
    4'b001? : y = 2'b01;
    4'b0001 : y = 2'b00;
    default : y = 2'b00;
endcase
```

By including the full_case attribute, the designer is telling the synthesis tool to treat any unspecified combinations of inputs as don't care conditions—in other words, to assume that a default item exists. Thus, the simulated and synthesized interpretations of the code would be different. Of course, if the default item is present, the full_case attribute is redundant!

The recommendation is, therefore, to omit the full_case attribute and to include a default item. The output values from the default should be meaningful values (not x or z), otherwise a latch might result. In the previous example, setting the default result to 2b'xx would mean that no valid hardware could be produced for the default cases and hence a latch would be implied. It is usual to include the default item as the last case item.

A parallel case statement is one in which each combination of inputs is covered exactly once. It is perfectly legal to write a case statement such that an input pattern can match to two or more case items. Because the items are matched in the order they are written, this implies the existence of priority logic. The use of a case statement suggests, however, that parallel logic should be used. Adding the parallel_case attribute forces the synthesis tool to treat the case statement as if it really is parallel. Inevitably, this will result in synthesized hardware that behaves differently from what was simulated at RTL.

The previous example is parallel. Including the parallel_case attribute is therefore redundant. The following example is not parallel:

```
always @(a)
  casez (a)
    4'b1??? : y = 2'b11;
    4'b?1?? : y = 2'b10;
    4'b??1? : y = 2'b01;
    4'b???1 : y = 2'b00;
    default : y = 2'b00;
  endcase
```

In a simulation, this code would appear to function in an identical manner to the priority encoder. The pattern 4'b1011 would match the first pattern, and so 2'b11 would result. This pattern also matches the third and fourth items. Clearly, therefore, the case statement is not parallel. By specifying the parallel_case attribute, the designer would be attempting to fool the synthesis tool—and would probably fail.

Therefore, do not use the parallel_case attribute. If you must have priority logic, use if statements. If you use case statements, read the messages from the synthesis tool. If the tool reports that your case statement is not parallel, *change the case statement* to make it parallel.

10.3 Synthesis for FPGAs

In principle, an RTL model of a piece of hardware coded in SystemVerilog can be synthesized to any target technology. In practice, the different technologies and structures of ASICs and FPGAs mean that certain constructs will be more efficiently synthesized than others and that some rewriting of SystemVerilog may be needed to attain the optimal use of a particular technology.

In this section we compare two FPGA technologies and show how the SystemVerilog coding of a design can affect its implementation in a technology. The descriptions of the technologies are deliberately simplified.

FPGAs are based on static RAM technology. Each FPGA consists of an array of the configurable logic blocks (CLBs) shown in Figure 1.13. Each logic block has one or more flip-flops and a combinational block. Each flip-flop has an asynchronous set and reset, but only one of these may be used at one time. Each flip-flop also has a clock input that can be positive or negative edge-sensitive, and each flip-flop has a clock enable input. In addition to the CLB shown, a number of three-state buffers exist in the array.

CPLDs are based on antifuse technology. Two types of logic block exist in more or less equal numbers—a combinational block and a sequential block. Each flip-flop in a sequential block has an asynchronous reset.

Both types of FPGA therefore have a relatively high ratio of flip-flops to combinational logic. Conventional logic design methods tend to assume that flip-flops are relatively expensive and combinational logic is relatively cheap, and that therefore sequential systems such as state machines should be designed with a minimal number of flip-flops. The large number of flip-flops in an FPGA and the fact that the flip-flops in an FPGA or a CPLD cannot be used without the combinational logic reverses that philosophy and suggests that one-hot encoding is a more efficient state encoding method, particularly for small state machines.

Similarly, a global asynchronous set or reset is the most efficient way of initializing both types of device. If both set and reset are required, it is necessary to use additional combinational logic, hence it may be better to have, for example, an asynchronous reset and a *synchronous* set.

In both technologies, the flip-flops are edge-sensitive; therefore, level-sensitive latches have to be synthesized from combinational logic. Again, this can waste flip-flops, so level-sensitive designs are best avoided. It is, however, reasonable to assume that any level-sensitive latches will exist as library elements and therefore that they will be hazard-free.

In both technologies, it may be desirable to instantiate predefined library components for certain functions. Not only is the logic defined, but the configuration of logic blocks is already known, potentially simplifying both the RTL synthesis and place and route tasks.

All the foregoing comments distinguish synthesis to FPGAs from synthesis to ASICs in general. The FPGA technologies themselves favor certain SystemVerilog coding styles. For example, the following piece of SystemVerilog shows two ways of describing a 5-to-1 multiplexer.

```
module Mux1(input a, b, c, d, e,
            input [4:0] s, output logic y);

always_comb
  case (s)
    5'b00001 : y = a;
    5'b00010 : y = b;
    5'b00100 : y = c;
    5'b01000 : y = d;
    default :  y = e;
  endcase
endmodule

module Mux2(input a, b, c, d, e,
            input [4:0] s, output wire y);

assign y = s[0] ? a : 1'bZ;
assign y = s[1] ? b : 1'bZ;
assign y = s[2] ? c : 1'bZ;
assign y = s[3] ? d : 1'bZ;
assign y = s[4] ? e : 1'bZ;
endmodule
```

These two models have the same functionality when simulated. If version 1 were synthesized to an FPGA, two CLBs would be needed. Version 2, on the other hand, can be implemented using the three-state buffers that exist outside the CLBs. Version 2, however, cannot be synthesized to a CPLD as the technology does not support three-state logic, except at the periphery. Clearly, therefore, the choice of architecture depends upon which technology is being used.

The two technologies have different limitations with respect to fan-outs. Antifuse technology has a fan-out limit of about 16 (one output can drive up to 16 inputs without degradation of the signal). CMOS SRAM technology has a higher fan-out limit. In practice, this means that a design that can be synthesized easily to a Xilinx FPGA cannot be synthesized to a CPLD without rewriting. For example, an apparently simple structure such as the following fragment cannot be synthesized as it stands because the Enable signal is controlling 32 multiplexers.

```
logic [31:0] a, b;

always_comb
  if (Enable)
    a = b;
  else
    a = 0;
```

Instead, the Enable signal must be split into two using buffers, and each buffered signal then controls half of the bus:

```
logic [31:0] a, b;
wire En0, En1;

buf b0 (Enable, En0);
buf b1 (Enable, En1);

always_comb
  begin
  if (En0)
    a[15:0] = b[15:0];
  else
    a[15:0] = 0;
  if (En1)
    a[31:16] = b[31:16];
  else
    a[31:16] = 0;
  end
```

A good synthesis tool should recognize the fan-out limits and automatically insert buffers.

10.4 Behavioral Synthesis

In RTL synthesis, the design is specified in terms of register operations and transformed automatically into gates and flip-flops. Behavioral synthesis takes the process one stage further. The hardware to be synthesized is described in terms of an algorithm, from which the registers and logic are derived. In principle, it is not necessary to use a hardware description language for behavioral synthesis; indeed, subsets of conventional programming languages such as C have been used. The major obstacle to the widespread acceptance of behavioral synthesis appears to be the difficulty that a hardware designer has in interpreting the output of a synthesis tool. The output of

RTL synthesis, particularly when expressed in terms of FPGA netlists, can be very difficult to interpret. This is even truer of behavioral synthesis, where the detailed structure is entirely generated by the synthesis tool. With the decreasing cost of silicon, however, it seems safe to predict that behavioral synthesis will become an accepted design tool, in the same way that compilers for high-level programming languages are now accepted, even though the machine code generated is largely unintelligible.

This section shows, by example, how a behavioral synthesis tool might generate a structural representation of a circuit from a high-level algorithmic description.

The following is a behavioral model of an infinite impulse response (IIR) filter.

```
module iir (input int in, output int out);

  const int coeffa [0:5] = '{25,50,75,150,300,600};
  const int coeffb [0:4] = '{-100,-125,-150,-175,-200};
  parameter order = 5;

  int input_sum = 0;
  int output_sum = 0;
  int delay [0:order] = '{0,0,0,0,0,0};
  always
    begin
    input_sum =in;
    for (int j = 0; j <= order-1; j++)
      input_sum = input_sum + (delay[j]*coeffb[j]/1024);
    output_sum = (input_sum*coeffa[order]/1024);
    for (int k = 0; k <= order; k++)
      output_sum = output_sum + (delay[k]*coeffa[k]/1024);
    for (int l = 0; l <= order-1; l++)
      delay[l] = delay [l+1];
    delay[order] = input_sum;
    out = output_sum ;
    #10ns;
    end

endmodule
```

This is a behavioral description in the sense that the filter is described purely as an algorithm. A C version of the algorithm would look very similar. A C version might not include the 10 ns delay, but conversely, this is not RTL SystemVerilog, as there is neither a clock nor a reset. If this description were used for RTL synthesis (assuming the synthesis tool accepted the SystemVerilog), the resulting hardware would have twelve 32-bit combinational multipliers and eleven 32-bit adders. This translates to 12,640 full adders. The division by 1024 is simply a scaling operation and can be achieved by throwing away the 10 least significant bits from each multiplication product. This operation is therefore effectively free.

The essential fact about behavioral synthesis is that it is possible to make design decisions and to achieve a compromise between speed and size. In the IIR example, it would be equally possible to implement the algorithm using 12,640 full adders and complete the operation in one clock cycle, or to use one full adder and take 12,640 clock cycles to achieve the result. More sensibly, some implementation between these two extremes might be sufficiently fast and sufficiently small to satisfy the requirements of the final application.

It is not practical to demonstrate the principles of behavioral synthesis with the fifth-order IIR filter. Instead, let us consider how a first-order filter might be built. In order to know which operations can be done concurrently and which require successive clock cycles, we need to know the dependency of each piece of data on each other piece of data. To do this, the loops in the behavioral description will first be expanded. We will ignore the division operations for the reason stated previously.

```
input_sum = in + delay[0]*coeffb[0];
output_sum = input_sum*coeffa[1];
output_sum = output_sum + delay[0]*coeffa[0];
out = output_sum + delay[1]*coeffa[1];
```

Assignments are made to `output_sum` on the second and third lines. To distinguish between successive values of `output_sum`, the two values will be separated, such that there is only one assignment to each variable in the algorithm. This is known as *single assignment form*.

```
input_sum = in + delay[0]*coeffb[0];
output_sum0 = input_sum*coeffa[1];
output_sum1 = output_sum0 + delay[0]*coeffa[0];
out = output_sum1 + delay[1]*coeffa[1];
```

From this a data dependency graph can be constructed (Figure 10.9).

If the operations shown in Figure 10.9 were all performed in one clock cycle, three adders and four multipliers would be needed. If it were decided, however, that each multiplication and each addition takes one clock cycle, the data dependency graph can be used to construct a *schedule* that shows when each operation can be performed (Figure 10.10).

It can be seen that five clock cycles are required to perform the arithmetic operations—the system is said to have a *latency* of five. This schedule is known as an *as soon as possible* (ASAP) schedule because each operation is done as early

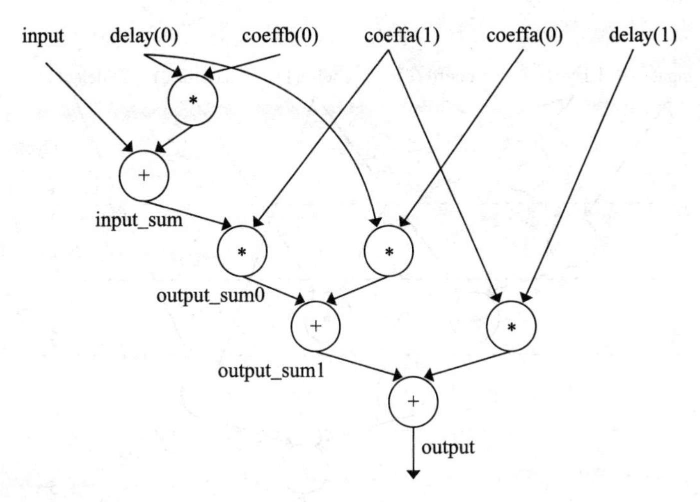

Figure 10.9 Data dependency graph.

as possible. Note that the sequence of operations is not the same as given by the original SystemVerilog description. Equally, it is possible to schedule operations *as late as possible* (ALAP) (Figure 10.11). This schedule also takes five clock cycles. If, however, the resources were constrained to a single arithmetic unit, again using an ALAP schedule, the number of cycles required increases (Figure 10.12).

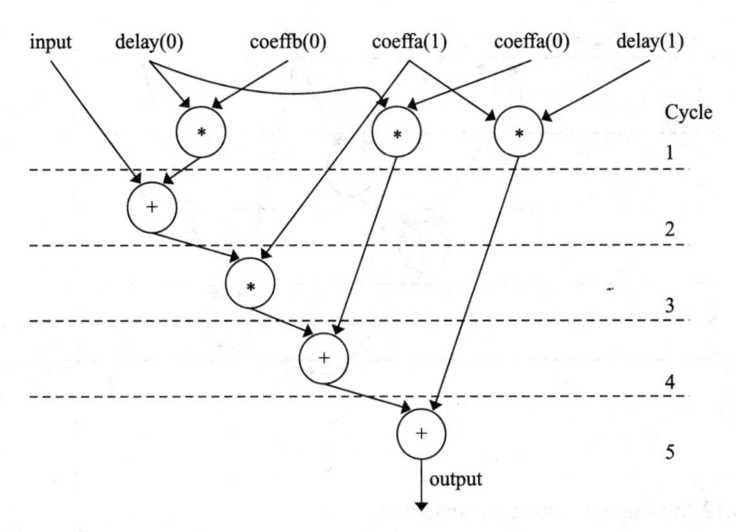

Figure 10.10 ASAP schedule.

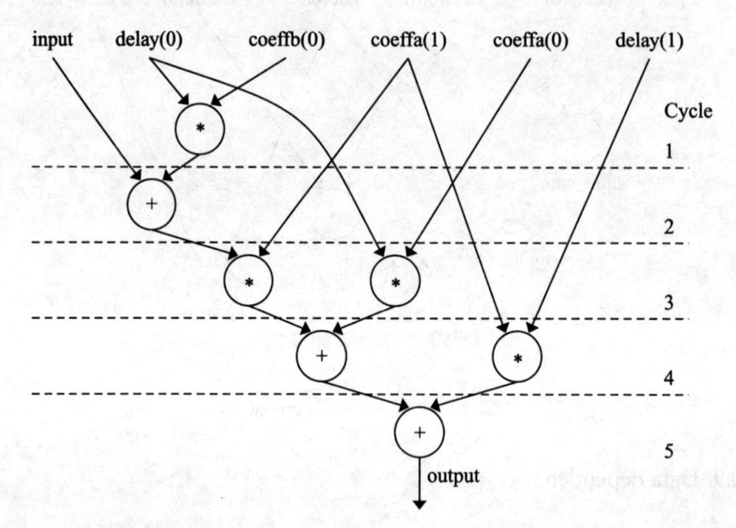

Figure 10.11 ALAP schedule.

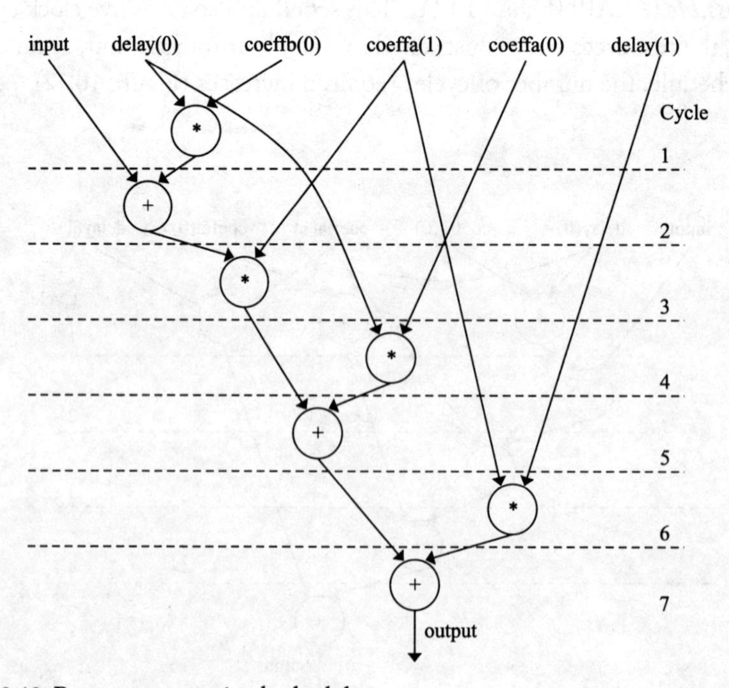

Figure 10.12 Resource constrained schedule.

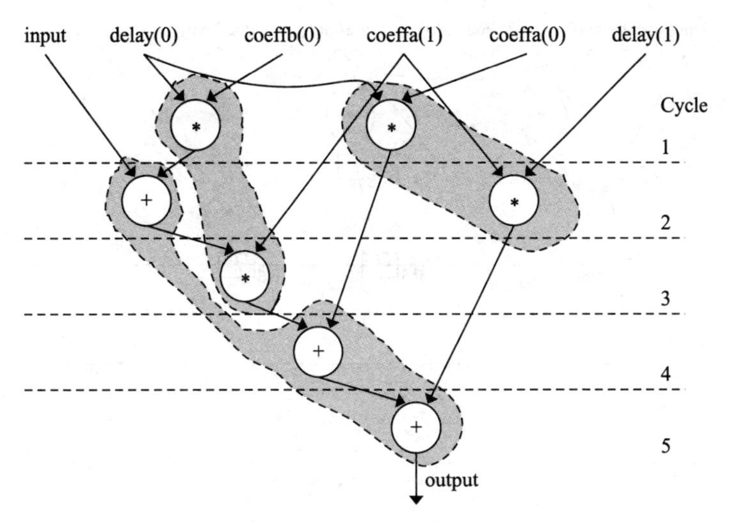

Figure 10.13 Mapping of operations onto resources.

Given the assumption that the basic resources available are arithmetic units, there are relatively few possible schedules for this example. With larger problems, the number of possible schedules clearly increases. By limiting the available resources, and hence the total area of the design, the latency, that is, the time taken to complete an operation, is increased. Therefore, the synthesis tool can trade speed against area by changing the schedule. Figure 10.13 shows how the operations can be mapped onto particular resources. The three shaded groups each represent, a resource used in different clock cycles, namely two multipliers and an adder.

The result of an operation is used in a subsequent clock cycle. Therefore, every time a data arc crosses a clock boundary, a register must be inserted, as shown in Figure 10.14.

Just as the arithmetic resources can be shared, so too can the registers be shared. The sharing is achieved using multiplexers, which are assumed to be cheap (i.e., small) compared with the other resources. Hence, a possible hardware implementation of the schedule of Figure 10.14 is shown in Figure 10.15.

In Figure 10.14 and Figure 10.15, three registers are shown following one of the multiplier units. This assumes that every register is loaded at each clock edge. It would be equally valid to use enabled registers, which would reduce the number of registers. Whatever technique is used, the multiplexers and registers have to be controlled. We have thus far discussed the derivation of the datapath part of Figure 7.6 from a behavioral description. The controller part also needs to be synthesized. In the example shown, this is relatively simple. There are five clock cycles; hence, five states as shown in Figure 10.16.

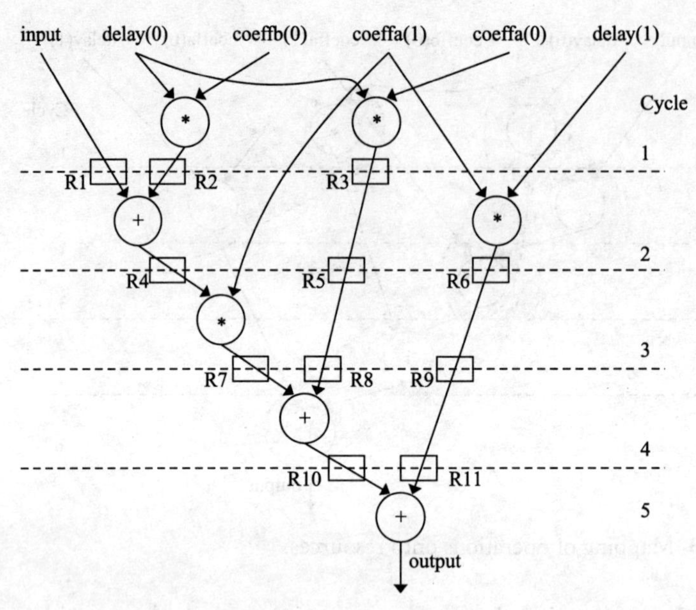

Figure 10.14 Schedule showing registers.

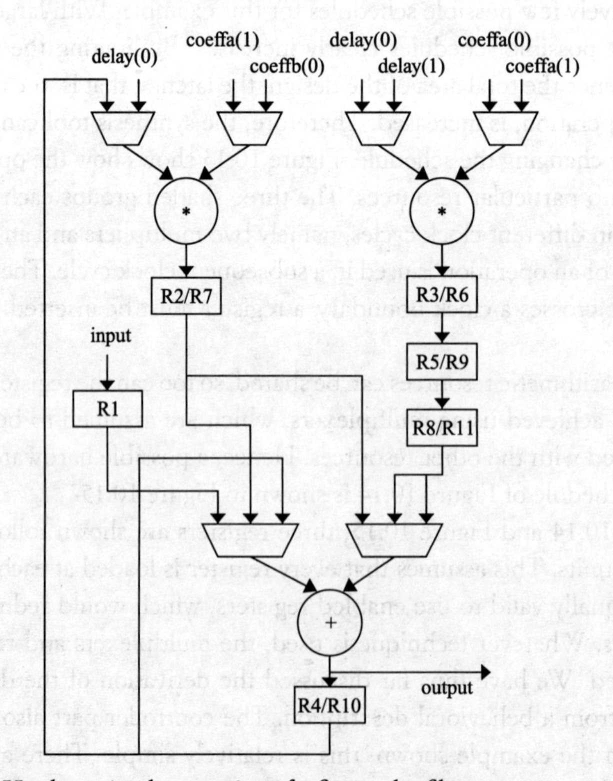

Figure 10.15 Hardware implementation of a first-order filter.

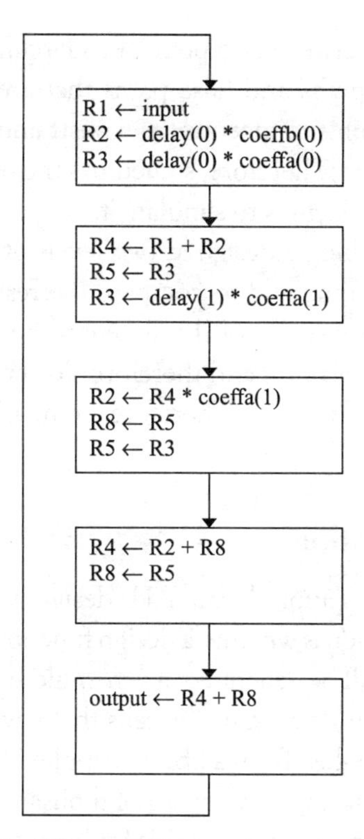

Figure 10.16 ASM chart of the IIR controller.

10.5 Verifying Synthesis Results

Synthesis should, by definition, produce a correct low-level implementation of a design from a more abstract description. In principle, therefore, functional verification of a design after synthesis should not be needed. For peace of mind, we might wish to check that the synthesized design really does perform the same function as the RTL description. Synthesis does, however, introduce an important extra factor to a design—timing. An RTL design is effectively cycle-based. A task takes a certain number of clock cycles to complete, but we do not really know how long each cycle takes. After synthesis, the design is realized in terms of gates or other functional blocks, and these can be modeled with delays. After placement and routing, we have further timing information in the form of wiring delays, which can be significant and which can affect the speed at which a design can operate.

It is possible, in principle, to verify a synthesized design by comparing it with the original RTL design, using techniques such as model-checking. In practice, such tools are limited to checking interfaces. Static timing analysis can give us information about delays between two points in a circuit, but needs to distinguish between realizable signal paths and false paths that are never enabled in reality. Similarly, a synthesis tool aims to meet timing constraints, but may not distinguish between true and false paths. Therefore, sometimes the best way to verify the timed behavior of a synthesized system is to simulate it.

One approach to checking a design at two levels of abstraction is to simulate both versions at the same time and to compare the results. This is usually a bad idea for two reasons. First, the size of the system to be simulated is at least twice as large as one version in isolation, and therefore slower to execute. Second, there will, as noted, be timing differences. Therefore, comparing responses may lead to false warnings.

10.5.1 Timing Simulation

The major concern for simulation during RTL design is to verify the functionality. The only timing consideration is whether a design functions correctly cycle-by-cycle. After synthesis, two files will be generated: a *netlist* file and a timing file in SDF. The netlist, as the name suggests, is a list of the cells that have been used to implement the design. These cells are taken from a library supplied by the FPGA vendor or by a silicon foundry. The cell library may consist of thousands of cells. Some are simply variants on a function that have a higher drive for higher fan-outs. The functionality of each cell can be expressed in terms of built-in gates or in terms of **primitive** modules. (We will not describe **primitive** modules here.) Normally, the netlist will have been "flattened"—all the hierarchy of the design will have been removed—so it is very difficult to understand the functionality of a design from the post-synthesis netlist.

The SDF timing file contains point-to-point timing information for the cells and the interconnect. This information can be extracted once the full placement and routing of the design is complete. The cells themselves will have been characterized by the builder of the cell library. SDF files contain three timing values for each path: fast, typical, and slow. In general, the slow values are used because they represent the worst-case behavior. Under some circumstances, fast values may also be simulated in order to check that flip-flop setup and hold times are not violated.

Within a netlist file, delay times are represented using **specify** blocks, not by the delay parameters of gate models. For example, the circuit of Figure 3.1 is modeled by the following netlist (reproduced from Chapter 3).

```
module ex1 (output wire y; input wire a, b, c);
  wire d, e, f;
  not g1 (d, c);
```

```
  and g2 (e, a, c);
  and g3 (f, d, b);
  or g4 (y, e, f);
endmodule
```

Instead of giving delay parameters to each gate, the delays between each input and the output might be given by a **specify** block as follows.

```
specify
  (a => y) = 9ns;
  (b => y) = 10ns;
  (c => y) = 12ns;
endspecify
```

(The **specify** block is included within the module.) When the delays are specified in an SDF file, the delay values are all set to zero in the netlist file.

```
specify
  (a => y) = 0;
  (b => y) = 0;
  (c => y) = 0;
endspecify
```

Thus, the timing simulation associates values from the SDF file with particular paths. At the time of simulation, fast, typical, or slow values are chosen for the *entire* circuit. We will look again at **specify** blocks in Chapter 13, when considering setup and hold times.

Static timing analysis (STA) uses the same timing information, but delays through entire paths are calculated from the cell and interconnect data. This is a little more complex than simply summing delays as the switching characteristics of gates have to be taken into account. While STA is much faster than timing simulation, care has to be taken with "false paths," as noted previously.

Whichever approach is used, the aim of timing analysis is to verify that a design will work at the intended clock speed. Thus, the analysis tends to be pessimistic. If a design does not meet its timing specification, it will have to be re-synthesized. As noted earlier in this chapter, this process is controlled using synthesis constraints. Optimizing a design to make it faster can result in its becoming larger. While this may decrease logic delays, wiring delays can increase. Therefore, it may be difficult to force a design to converge to acceptable timing constraints. Timing simulation and STA clearly play a big role in this cycle and therefore have to be efficient.

Summary

SystemVerilog was conceived as a description language, but has been widely adopted as a specification language for automatic hardware synthesis. A number of tools exist for RTL synthesis, but behavioral synthesis tools are appearing. Because of its origins, SystemVerilog has some features that are not synthesizable to hardware. The rules for the inference of latches and flip-flops are well defined. Synthesis constraints may be stated in terms of SystemVerilog attributes or as separate inputs to the synthesis tool. To get the most out of an FPGA may require careful writing of the SystemVerilog code. The important concepts behind behavioral synthesis are scheduling and binding. Verification of synthesized designs can be done by comparing pre- and post-synthesis simulations. The detailed timing of designs can be modeled using timing simulation or STA.

Further Reading

Despite the definition of a synthesizable subset of Verilog, each synthesis tool accepts a slightly different subset of Verilog and SystemVerilog and interprets poorly written SystemVerilog in different ways. It therefore pays to read the user manuals of tools with some care. The Web sites of FPGA manufacturers include SystemVerilog style guides showing what can and cannot be implemented.

de Micheli [6] covers both high-level behavioral synthesis and low-level optimization in his book.

Exercises

10.1 Explain, with examples, what is meant by a constraint in RTL synthesis.

10.2 Write a model of an eight-state counter as a SystemVerilog state machine, with a clock and reset inputs, which outputs a ready flag when the counter is in the initial state. Use the `enum_encoding` attribute to specify that the state machine should be implemented as a Johnson counter.

10.3 The following listing shows a description of a simple state machine in SystemVerilog. If this state machine were synthesized using an RTL synthesis tool, the resulting hardware would give different simulated behavior from the original RTL description. Explain why this should be so.

```
module fsm (input logic clk, a, output logic y);

  typedef enum {s0, s1, s2} statetype;
```

```
  statetype currentstate, nextstate = s0;

always @(posedge clock)
  currentstate <= nextstate;

always @(currentstate)
  case (currentstate)
    s0: if (a)
          nextstate = s1;
        else
          nextstate = s2;
    s1: begin
        y = '1;
        nextstate = s0;
        end
    s2: if (a)
          nextstate = s2;
        else
          nextstate = s0;
    endcase

endmodule
```

10.4 Rewrite the SystemVerilog model of Exercise 10.3 such that, when synthesized, the resulting hardware consists only of D flip-flops, with asynchronous resets and combinational next state and output logic.

10.5 The following listing shows three SystemVerilog processes. Describe the hardware that should be generated from each of these process models by a synthesis tool.

```
always @(x, y)
begin: A
  if (y)
    qa = x;
  else
    qa = '0;
end

always @(x, y)
begin: B
  if (y)
    qb = x;
end

always @(y)
```

```
begin: C
  if (y)
    qc = x;
end
```

10.6 Explain the terms *scheduling* and *binding* in the context of behavioral synthesis.

10.7 The following sequence of operations is part of a cube root solution routine:

```
a = x * x;
a = 3 * a;
b = y / a;
a = 2 * x;
a = a / 3;
c = a - b;
```

Convert this sequence to single assignment form and hence construct a data dependency graph. Assuming that each arithmetic operation takes exactly one clock cycle, derive an unconstrained ALAP schedule.

10.8 Derive a constrained schedule for the routine of Exercise 10.7 and hence design a datapath implementation of this part of the system, assuming that one multiplier, one divider, and one subtracter are available.

11 Testing Digital Systems

In the course of manufacture, defects may be introduced into electronic systems. Systems may also break during use. Defects may not be easy to detect. In this chapter we discuss the importance of testing, the types of defects that can occur, and how defects can be detected. We describe procedures for generating tests and how the effectiveness of tests can be assessed.

11.1 The Need for Testing

No manufacturing process can ever be perfect. Thus, real electronic systems may have manufacturing defects such as short circuits, missing components, or damaged components. A manufacturer needs to know if a system (whether at the level of a board, an IC, or a whole system) has a defect and therefore does not work in some way. While a manufacturer does not want to sell bad systems, equally he or she would not want to reject good systems. Therefore, the need for testing is economic.

We also need to distinguish between the ideas of *verification* in which the design of a piece of hardware or software is checked and of *testing* in which it is assumed that the design is correct, but that there may be manufacturing faults. This chapter is about the latter concept, but the inclusion of design for test structures *may* help in verifying and debugging a design.

There are, in general, two approaches to testing. We can ask whether the system works correctly (*functional* testing) or we can ask whether the system contains a fault (*structural* testing). These two approaches might at first appear to be equivalent, but in fact the tactic we adopt can make a profound difference to how we develop tests and how long those tests take to apply. Functional testing can imply a long and difficult task because all possible states of a system have to be checked. Structural testing is often easier, but is dependent upon the exact implementation of a system.

11.2 Fault Models

An electronic system might contain a large number of possible defects as a result of the manufacturing process. For example, the printed circuit board could have breaks in connections because of bad etching, stress, or bad solder joints. Equally, there may be short circuits resulting from the flow of solder. The components on a printed circuit board (PCB) may be at fault—so-called "population defects"—caused by having the wrong components, wrongly inserted components, or omitted components. The components themselves may fail because the operating conditions exceed the component specifications or because of electromagnetic interference (EMI) or heat.

Similar defects can occur in integrated circuits. Open circuits may arise from electromigration (movement of metal atoms in electromagnetic fields), current overstress, or corrosion. Silicon or oxide defects, mask misalignment, impurities, and gamma radiation can cause short circuits and incorrect transistor operation. "Latch-up," caused by transient currents, forces the output of a CMOS gate to be stuck at a logic value. In memory circuits, there may be data corruption because of alpha particles or EMI.

Clearly, to enumerate and check for every possible defect in an electronic system would be an enormous task. Therefore, a distinction is made between physical *defects* and electrical *faults*. The principle of fault modeling is to reduce the number of effects to be tested by considering how defects manifest themselves. A physical defect will manifest itself as a logical fault. This fault may be static (e.g., shorts, breaks), dynamic (components out of specification, timing failures), or intermittent (environmental factors).

The relative probabilities of faults that appear during tests in manufacturing are shown in Figure 11.1. Dynamic faults may be further divided into timing faults (28%) and driver faults (21%). Timing faults and intermittent faults may be due to poor design. It is difficult to design test strategies for such faults.

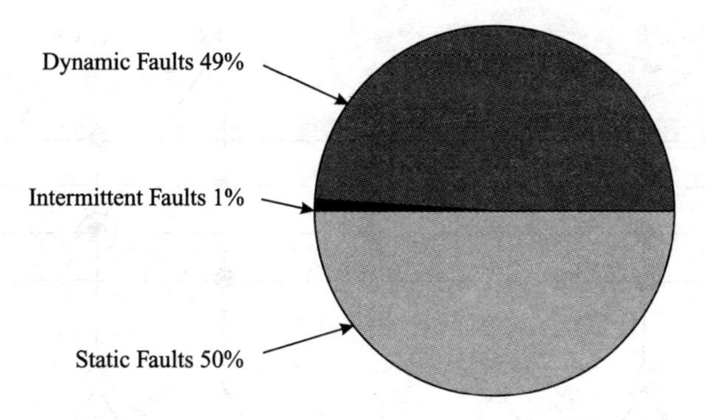

Figure 11.1 Fault probabilities.

11.2.1 Single-Stuck Fault Model

Static faults are usually modeled by the *stuck fault model.* Many physical defects can be modeled as a circuit node being either stuck at 1 (s-a-1) or stuck at 0 (s-a-0). Other fault models include stuck open and stuck short faults. Programmable logic and memory have other fault models.

The *single-stuck fault model* (SSFM) assumes that a fault directly affects only one node and that the node is stuck at either 0 or 1. These assumptions make test pattern generation easier, but the validity of the model is questionable. Multiple faults do occur, and multiple faults can theoretically mask each other. On the other hand, the model appears to be valid most of the time. Hence, almost all test pattern generation relies on this model. Multiple faults are generally found with test patterns for single faults.

11.2.2 PLA Faults

PLAs consist not of gates but of AND and OR logic planes, connected by fuses (or anti-fuses). Thus, faults are likely to consist of added or missing fuses, not stuck faults. For example, Figure 11.2 shows part of a PLA, where the output Z is the logical OR of three intermediate terms, P, Q, R.

Each of the intermediate terms is the AND of the three inputs, A, B, C or its inverse:

$$
\begin{aligned}
Z &= P + Q + R \\
P &= B \cdot \bar{C} \\
Q &= A \cdot C \\
R &= \bar{A} \cdot \bar{B} \cdot \bar{C} \\
S &= \bar{A} \cdot C
\end{aligned}
$$

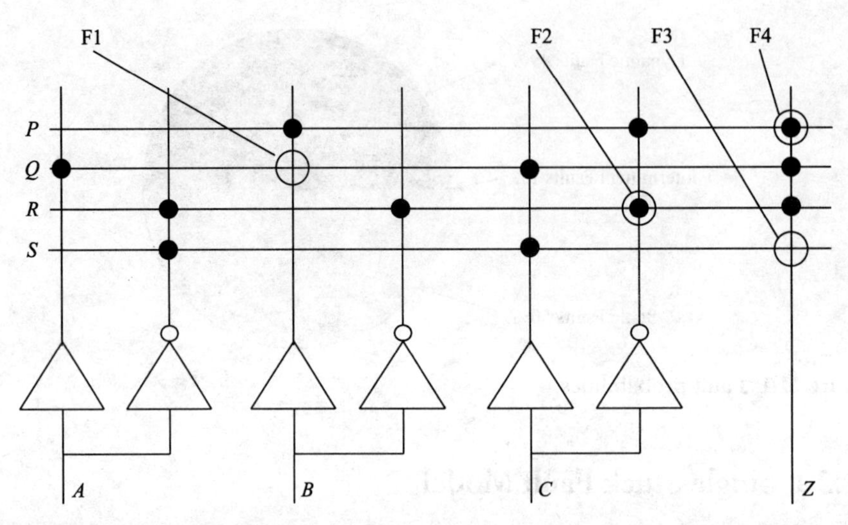

Figure 11.2 PLA fault models.

- Fault F1 is an additional connection causing Q to change from $A \cdot C$ to $A \cdot B \cdot C$. On a Karnaugh map this represents a decrease in the number of 1s circled; therefore, this can be thought of as a *shrinkage* fault.
- Fault F2 is a missing connection, causing R to *grow* from $\bar{A} \cdot \bar{B} \cdot \bar{C}$ to $\bar{A} \cdot \bar{B}$.
- Fault F3 causes the *appearance* of term S in Z.
- Fault F4 causes the *disappearance* of term P from Z.

11.3 Fault-Oriented Test Pattern Generation

Having decided that defects in a system can be modeled as electrical faults, we then need to determine whether any of these faults exist in a particular instance of a manufactured circuit. If the circuit were built from discrete transistors or gates, this task could, in theory, be achieved by monitoring the state of every node of the circuit. If the system is implemented as a packaged integrated circuit, this approach is not practical. We can only observe the outputs of the system, and we can only control the inputs of the system. Therefore, the task of test pattern generation is that of determining a set of inputs to unambiguously indicate if an internal node is faulty. If we only consider combinational circuits for the moment, the number of possible input combinations for an n-input circuit is 2^n. We could apply all 2^n inputs, in other words, perform an exhaustive functional test, but in general we want to find the minimum necessary number of input patterns. It is possible that, because of the circuit structure, certain faults cannot be detected. Therefore, it is common to talk about the *testability* of a circuit.

Testability can be a somewhat abstract concept. One useful definition of testability breaks the problem into two parts:

1. *Controllability*—Can we control all the nodes to establish if there is a fault?
2. *Observability*—Can we observe and distinguish between the behavior of a faulty node and that of a fault-free node?

In order to generate a minimum number of test patterns, a fault-oriented test generation strategy is adopted. In the pseudo-code that follows, a *test* is one set of inputs to a (combinational) circuit. The overall strategy is as follows:

- Prepare a fault list (e.g., all nodes stuck-at 0 and stuck-at 1)
- Repeat
  - Write a test
  - Check fault cover (one test may cover more than one fault)
  - Delete covered faults from list
- until the fault cover target is reached

Test pattern generation (writing a test) may be random or optimized. This will be discussed in more detail later. One test may cover more than one fault, and often faults are indistinguishable. Again, this is discussed later.

If we simply want a pass/fail test, once we have found a test for a fault we can remove faults from further consideration. If we want to diagnose a fault (for subsequent repair), we probably want to find all tests for a fault to deduce where the fault occurs. The fault cover target may be less than 100%. For large circuits, the time taken to find all possible tests may be excessive. Moreover, the higher the cover, the greater the number of tests and hence the cost of applying the test.

11.3.1 Sensitive Path Algorithm

The circuit of Figure 11.3 has seven nodes. Therefore, there are 14 stuck faults:

$$A/0,\ A/1,\ B/0,\ B/1,\ C/0,\ C/1,\ D/0,\ D/1,\ E/0,\ E/1,\ F/0,\ F/1,\ Z/0,\ Z/1$$

where $A/0$ means A stuck-at-0, etc.

To test for $A/0$, we need to set A to 1 (the fault-free condition—if A were at 0, we would not be able to distinguish the faulty condition from the fault-free state). The presence or otherwise of this fault can only be detected by observing node Z.

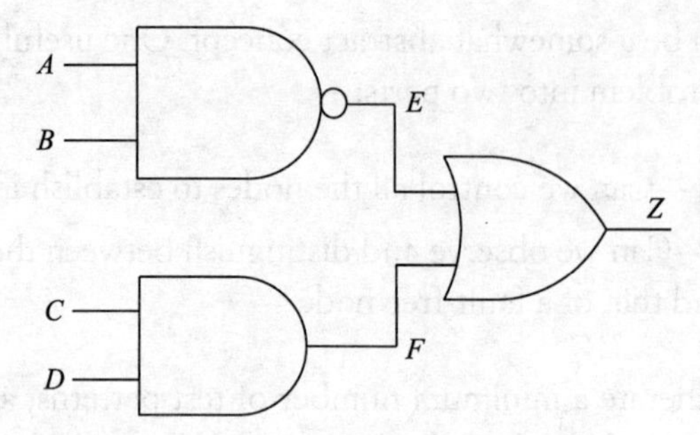

Figure 11.3 Example circuit for test generation.

We now have to determine the states of the other nodes of the circuit that allow the state of A to be deduced from the state of Z. Thus, we must establish a sensitive path from A to Z. If node B is 0, E is 1 irrespective of the state of A. Therefore, B must be set to a logic 1. Similarly if F is 1, Z is 1, irrespective of E; hence, F must be 0. To force F to 0, either C or D or both must be 0.

Thus, if the fault $A/0$ exists, E is 1 and Z is 1. If the fault does not exist, E is 0 and Z is 0.

We can conclude from this that a test for $A/0$ is $A = 1$, $B = 1$, $C = 0$, $D = 1$, for which the fault-free output is $Z = 0$. This can be expressed as 1101/0. Other tests for $A/0$ are 1110/0 and 1100/0. Therefore, there is more than one test for the fault $A/0$.

Let us now consider a test for another fault. To test for $E/1$ requires that $F = 0$ to make E visible at Z. Therefore, C or D or both must be 0. To make $E = 0$ requires that $A = B = 1$. So, a test for $E/1$ is 1101/0. This is the same test as for $A/0$. So, one test can cover more than one fault.

The sensitive path algorithm therefore consists of the following steps.

1. Select a fault.
2. Set up the inputs to force the node to a fixed value.
3. Set up the inputs to transmit the node value to an output.
4. Check that the input node values for steps 2 and 3 are consistent.
5. Check for coverage of other faults.

The aim is to find the minimum number of tests that cover all the possible faults, although 100% fault coverage may not be possible.

Fan-out and reconvergence can cause difficulties for this algorithm. Improved algorithms (D-algorithm, PODEM) use similar techniques but overcome these drawbacks.

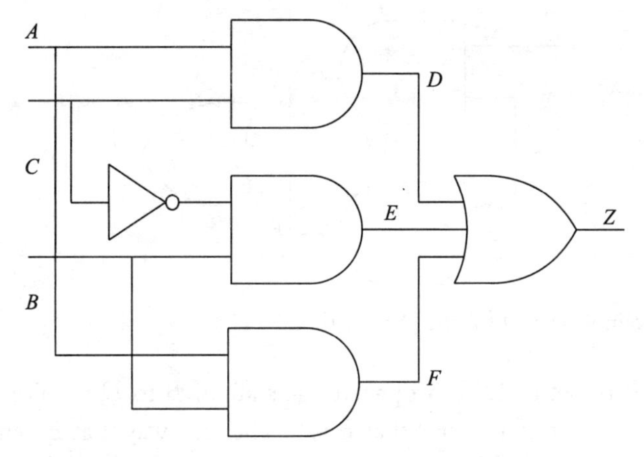

Figure 11.4 Circuit with redundancy.

11.3.2 Undetectable Faults

Consider the function

$$Z = A \cdot C + B \cdot \bar{C}$$

To avoid hazards, the redundant term may be included, as shown in Figure 11.4. This is effectively the same as Figure 2.17.

$$Z = A \cdot C + B \cdot \bar{C} + A \cdot B$$

We will now try to find a test for *F*/0. This requires that F be set to 1. Hence, $A = B = 1$. To transmit the value of F to Z means that $D = E = 0$ (otherwise Z would be 1, irrespective of F). For E to be 0, B must be 0 and/or C must be 1. Similarly, for D to be 0, A must be 0 and/or C must be 0. These three conditions are inconsistent, so no test can be derived for the fault *F*/0.

There are three possible responses to this. It must be accepted that the circuit is not 100% testable; the redundant gate must be removed, risking a hazard; or the circuit must be modified to provide a control input for testing purposes to force D to 0 when $A = C = 1$.

In general, untestable faults are due to redundancy. Conversely, redundancy in combinational circuits will mean that those circuits are not fully testable.

11.3.3 The D Algorithm

The simple sensitized path procedure does not handle reconvergent paths adequately. For example, consider the circuit of Figure 11.5. To find a test for *B*/0

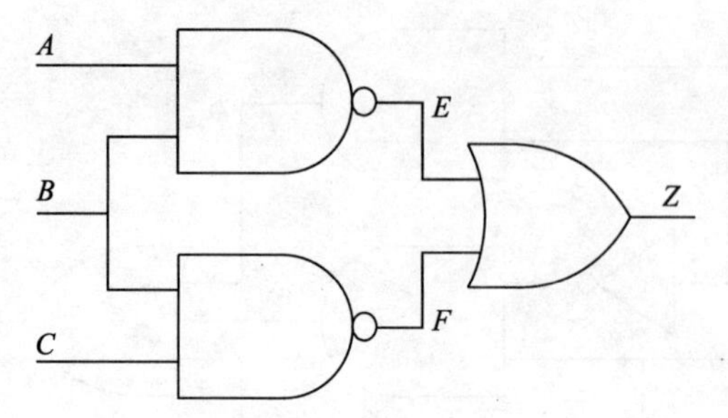

Figure 11.5 Example circuit for the D algorithm.

requires that B be set to 1. To propagate the state of B to D requires that A is 1, and to propagate D to Z requires that E is 0. The only way that E can be at 0 is if B and C are both 1, but this is not the case when $B/0$. Apparently, therefore, the sensitive path algorithm cannot find a test for $B/0$. In fact, 111/0 is a suitable test because under fault-free conditions, D, E, and Z are all at logic 0; when $B/0$, all three nodes are at logic 1.

The D algorithm overcomes that problem by introducing a 5-valued algebra: $\{0, 1, D, \bar{D}, X\}$. D represents a node that is logic 1 under fault-free (normal) conditions and logic 0 under faulty conditions. $\bar{D}$ represents a normal 0, and a faulty 1. X is an unknown value. The values of D and $\bar{D}$ are used to represent the state of a node where there is a fault and also the state of any other nodes affected by the fault.

The D algorithm works in the same way as the sensitive path algorithm. If step 4 fails, the algorithm backtracks. In both steps 2 and 3, it is possible that more than one combination of inputs generates the required node values. If necessary, all possible combinations of inputs are examined.

Table 11.1 shows the inputs required to establish a fault at an internal node, to transmit that fault to an output, to generate a fixed value (to establish or propagate a fault), and finally how fault conditions can reconverge. In all cases, the inputs A and B are interchangeable. The table can be extended to gates with three or more inputs. The symbol "–" represents a "don't care" input.

To see how the D notation can be used, consider the circuit of Figure 11.6. To test for $A/0$, node A is first given a value D, which can be propagated via node H or node G. To propagate the D to node H, node B must be 1. Node I then has the value $\bar{D}$. To propagate this $\bar{D}$ to K requires F to be 0 and to propagate the value to Z means J must be 1. If F is 0 and J is 1, G must be 1. Therefore, nodes A and E must both be 1. At this point we hit an inconsistency as node A has the value D. We have to return to the last decision made, which in this case was the decision to propagate the value of A through to H.

Table 11.1 Truth Tables for the D Algorithm

	AND			OR			NAND			NOR			NOT	
	A	B	Z	A	B	Z	A	B	Z	A	B	Z	A	Z
Establish fault-sensitive condition	1	1	D	0	0	$\bar{D}$	1	1	$\bar{D}$	0	0	D	1	$\bar{D}$
	0	–	$\bar{D}$	1	–	D	0	–	D	1	–	$\bar{D}$	0	D
Transmit fault	D	1	D	D	0	D	D	1	$\bar{D}$	D	0	$\bar{D}$	D	$\bar{D}$
	$\bar{D}$	1	$\bar{D}$	$\bar{D}$	0	$\bar{D}$	$\bar{D}$	1	D	$\bar{D}$	0	D	$\bar{D}$	D
Generate fixed value	1	1	1	1	–	1	1	1	0	1	–	0	1	0
	0	–	0	0	0	0	0	–	1	0	0	1	0	1
Reconvergence	D	D	D	D	D	D	D	D	$\bar{D}$	D	D	$\bar{D}$		
	$\bar{D}$	$\bar{D}$	$\bar{D}$	$\bar{D}$	$\bar{D}$	$\bar{D}$	$\bar{D}$	$\bar{D}$	D	$\bar{D}$	$\bar{D}$	D		
	D	$\bar{D}$	0	D	$\bar{D}$	1	D	$\bar{D}$	1	D	$\bar{D}$	0		

The alternative is to propagate the D at A to G. Thus, E must be 1. To propagate the value to J, F must be 0 and to propagate to Z, K must be 1. Hence, I must be 1, hence E must be 0. As A is already assigned, B must be 0. This is consistent with F being 0 and C may be either 1 or 0.

The D algorithm, as presented here, requires further refinement before it can be implemented as an EDA program. In particular, the rules for detecting inconsistencies require more detail. Table 11.2 shows what happens when two fault-free or faulty values are propagated by different routes to the same node.

The D algorithm is an algorithm in the true sense of the word—if a solution exists, the D algorithm will find it. The search for a solution can, however, be

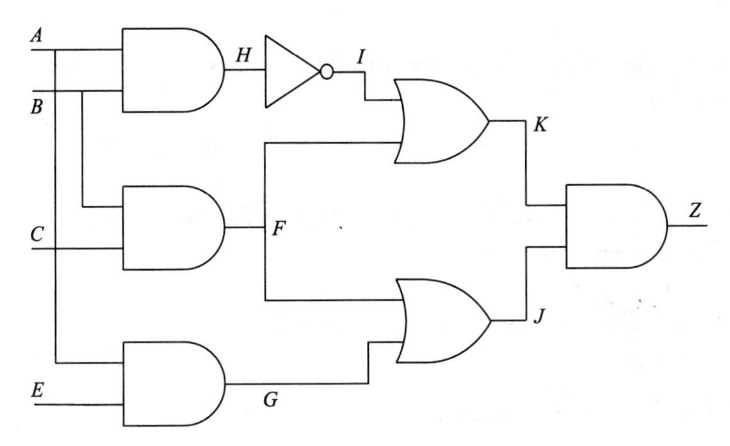

Figure 11.6 Example circuit for the D algorithm.

Table 11.2 Intersection Rules for the D Algorithm

$\cap$	0	1	X	D	$\bar{D}$
0	0	ϕ	0	ψ	ψ
1	ϕ	1	1	ψ	ψ
X	0	1	X	D	$\bar{D}$
D	ψ	ψ	D	μ	λ
$\bar{D}$	ψ	ψ	$\bar{D}$	λ	μ

ϕ = inconsistent logic values
ψ = inconsistency between logic values and fault values
μ = allowed intersection between fault values
λ = inconsistent fault values

very time-consuming. If necessary, every possible combination of node values will be examined. Subsequent test pattern generation algorithms have attempted to speed up the D algorithm by improving the decision making within the algorithm. Examples include 9-V which uses a nine-valued algebra and PODEM.

11.3.4 PODEM

The PODEM algorithm attempts to limit the decision making, and hence the time needed for a decision. Initially, all the inputs are set to X (unknown). Arbitrary values are then assigned to the inputs, and the implications of these assignments are propagated forward. If either of the following propositions is true, the assignment is rejected:

1. The node value of the fault under consideration has identical faulty and fault-free values.
2. There is no signal path from a net with a D or $\bar{D}$ value to a primary output.

We will use PODEM on the circuit of Figure 11.6 to develop a test for $I/1$. Initially, all nodes have an X value.

1. Set $A = 0$. Fails—proposition 1 (I would be 1).
2. Set $A = 1$. OK.
3. Set $B = 0$. Fails—proposition 1.
4. Set $B = 1$. OK. $H = D$, $I = \bar{D}$.

5. Set $C = 0$. OK. $F = 0$, $K = \bar{D}$.
6. Set $E = 0$. Fails—proposition 2 ($G = 0$, $J = 0$, $Z = 0$).
7. Set $E = 1$. OK. $G = 1$, $J = 1$, $Z = \bar{D}$.

Therefore, a test for $I/1$ is 1101/0

11.3.5 Fault Collapsing

In the example of Figure 11.3, the test for $A/0$ (the input to a NAND gate) was the same as the test for $E/1$ (the output of that NAND gate). The same test can be used to detect $B/0$. These three faults are *indistinguishable* $\{A/0, B/0, E/0\}$. Similarly, a test for an input of a NAND gate being stuck at 1 will also detect if the output is stuck at 0. Two different tests are needed, however, for $A/1$ and $B/1$. Hence, these faults are not indistinguishable, but an input stuck at 1 is said to *dominate* the output stuck at 0 ($A/1 \rightarrow E/0$). The set of rules for fault indistinguishability and dominance for two input (A, B), single output (Z) gates and the inverter are shown in Table 11.3.

These rules can be used to reduce a fault list. These rules, however, do not apply to fan-out nodes, which must be omitted from any simplification procedure. If we apply these rules to the 14 faults of the circuit of Figure 11.3, we can see that we have two sets of equivalent faults: $\{A/0, B/0, E/1, F/1, Z/1\}$ and $\{C/0, D/0, F/0\}$ and the following fault dominances: $A/1 \rightarrow E/0$, $B/1 \rightarrow E/0$, $E/0 \rightarrow Z/0$, $F/0 \rightarrow Z/0$, $C/1 \rightarrow F/1$, $D/1 \rightarrow F/1$. As we only need to test for one fault in each equivalent set and for the dominant faults, we only need to derive tests for the following faults: $A/1$, $B/1$, $C/1$, $D/1$, and $C/0$. The fault list is cut from 14 to 5 faults, simplifying the fault generation task. Note that we have not lost any information by doing this—we cannot tell by observing node Z whether a fault in the circuit is one of the five listed or a fault equivalent to or dominated by one of those faults.

Table 11.3 Fault Collapsing Rules

Type of Gate	Indistinguishable Faults	Fault Dominance
AND	$\{A/0, B/0, Z/0\}$	$A/1, B/1 \rightarrow Z/1$
OR	$\{A/1, B/1, Z/1\}$	$A/0, B/0 \rightarrow Z/0$
NAND	$\{A/0, B/0, Z/1\}$	$A/1, B/1 \rightarrow Z/0$
NOR	$\{A/1, B/1, Z/0\}$	$A/0, B/0 \rightarrow Z/1$
NOT	$\{A/0, Z/1\}$ $\{A/1, Z/0\}$	

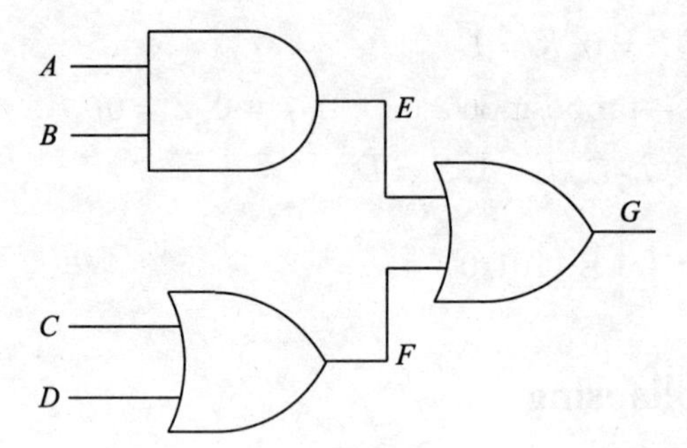

Figure 11.7 Example circuit for fault simulation.

11.4 Fault Simulation

One test pattern can be used to find more than one potential fault. For example, suppose we wish to detect if node E is stuck at 0 in the circuit of Figure 11.7. $E/0$ dominates $G/0$ and is equivalent to $A/0$ and $B/0$. In all these cases, G will be 1 normally and 0 in the presence of one of these faults. Hence, the input pattern $A = 1$, $B = 1$, $C = 0$, $D = 0$ can be used to detect four possible faults. As there are seven nodes in the circuit, there are 14 possible stuck-at faults. This pattern covers four faults, and it can be shown that of the 16 possible input patterns, six are sufficient to detect all the possible stuck-at faults in the circuit.

It is also generally true that a fault may be covered by more than one pattern. For instance, $E/1$ can be found by attempting to force E to 0. This can be achieved by setting (a) $A = 1$, $B = 0$, (b) $A = 0$, $B = 1$, or (c) $A = 0$, $B = 0$; in all cases, $C = 0$, $D = 0$. Thus, there are three possible patterns for detecting $E/1$. Note too that pattern (a) also covers $B/1$ and $G/1$, (b) covers $A/1$ and $G/1$, and (c) covers $G/1$. To detect all the faults in the circuit we need to use both $A = 1$, $B = 0$, $C = 0$, $D = 0$ and $A = 0$, $B = 1$, $C = 0$, $D = 0$ as these are the only patterns that detect $B/1$ and $A/1$, respectively. We are, however, applying two patterns that can detect $E/1$ and $G/1$. Having found one pattern that detects these two faults, we can *drop* the faults from further consideration. In other words, in applying the second test $A = 0$, $B = 1$, $C = 0$, $D = 0$, we forget about $E/1$ and $G/1$ as we already have a pattern that detects them. We could equally decide not to drop a fault when a suitable test pattern is found, in order to try to distinguish between apparently equivalent faults.

The object of fault simulation is, therefore, to assess the fault coverage of test patterns by determining whether the presence of a fault would cause the outputs of the circuit to differ from the fault-free outputs, given a particular input pattern.

The simplest approach to fault simulation is simply to modify the circuit to include each fault, one at a time, and to re-simulate the entire circuit. As the single-stuck fault model assumes that only one fault can occur at a time and that each node of the circuit can be stuck at 1 and at 0, this approach, known as *serial* fault simulation, will require twice as many simulation runs as there are nodes, together with one simulation for the fault-free circuit. This technique is clearly expensive in terms of computer power and time, and three main alternatives have been suggested to make fault simulation more efficient. We will show how two of these approaches can be implemented in a simulator.

11.4.1 Parallel Fault Simulation

If we use two-state logic, one bit is sufficient to represent the state of a node. Therefore, one computer word can represent the state of several nodes or the state of one node under several faulty conditions. For instance, a computer with a 32-bit word length can use one word to represent the state of a node in the fault-free circuit together with the state of the node when 31 different faults are simulated. Each bit corresponds to the circuit with one fault present. The same bit is used in each word to represent the same version of the circuit. The fault-free circuit must always be simulated as it is important to know whether a faulty circuit can be distinguished from the fault-free circuit. If more faults are to be simulated than the number of bits in a word, the fault simulation must be completed in several passes, each of which includes the fault-free circuit.

Instead of simulating the circuit by passing Boolean values, words are used, so the state of each gate is evaluated for each fault modeled by a bit of the input signal words; hence, the name *parallel fault simulation.* Because words are passed instead of Boolean values, the event scheduling algorithm treats any change in a word value as an event. Thus, gates may be evaluated for certain versions of the circuit even if the input values for that version remain unchanged.

The circuit of Figure 11.7 has seven nodes, and hence 14 possible stuck-at faults (Table 11.4). Thus, 15 bits are needed for a parallel fault simulation. The word values of each node for the input pattern $A = 1$, $B = 1$, $C = 0$, $D = 0$ are shown later. As can be seen, this pattern, as noted earlier, normally sets G to 1, but for faults $A/0$, $B/0$, $E/0$, and $G/0$, the output is 0, and therefore these faults are detected by that pattern.

There are several obvious disadvantages to parallel fault simulation. First, the number of faults that can be simulated in parallel is limited to the number of bits in a word. If more than two states are used—in other words if a state is encoded using two or more bits—the possible number of parallel faults is further reduced.

Table 11.4 Parallel Fault Simulation of the Circuit of Figure 11.7

Bit		A	B	C	D	E	F	G
0	–	1	1	0	0	1	0	1
1	A/0	0	1	0	0	0	0	0
2	A/1	1	1	0	0	1	0	1
3	B/0	1	0	0	0	0	0	0
4	B/1	1	1	0	0	1	0	1
5	C/0	1	1	0	0	1	0	1
6	C/1	1	1	1	0	1	1	1
7	D/0	1	1	0	0	1	0	1
8	D/1	1	1	0	1	1	1	1
9	E/0	1	1	0	0	0	0	0
10	E/1	1	1	0	0	1	0	1
11	F/0	1	1	0	0	1	0	1
12	F/1	1	1	0	0	1	1	1
13	G/0	1	1	0	0	1	0	0
14	G/1	1	1	0	0	1	0	1

As has been noted, every version of a gate is scheduled and re-evaluated whenever one of the versions of an input changes. This can be very inefficient as a significant number of null events are likely to be processed. Moreover, if the purpose of the fault simulation is simply to detect whether any of the given test patterns will detect any of the faults, it is desirable to drop a fault from further consideration once it has proved possible to distinguish the behavior caused by that fault from the normal, fault-free behavior. Faults cannot be dropped in parallel fault simulation, or perhaps more accurately, the dropping of a fault is unlikely to improve the efficiency of the simulation as the bits corresponding to that fault cannot be used for any other purpose.

11.4.2 Concurrent Fault Simulation

If only the differences between the fault-free simulation and the faulty simulations are maintained, constraints such as word size need not apply. On the other hand, the evaluation of gates would be made more complex because these lists of differences must be manipulated. Concurrent fault simulation maintains fault lists, in the form of those gates that have different inputs and outputs in the faulty circuit from the equivalent gates in the fault-free circuit. The manipulation of fault lists thus consists of evaluating input signals, in exactly the same way as is done for the fault-free circuit, and checking to see if the output differs from the fault-free circuit.

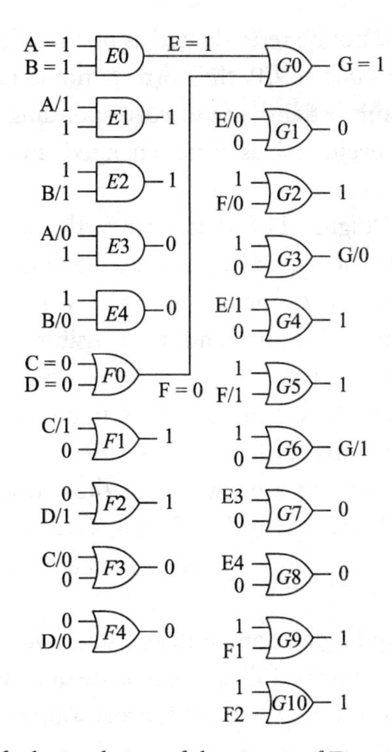

Figure 11.8 Concurrent fault simulation of the circuit of Figure 11.7.

Figure 11.8 shows the circuit with the fault lists included for the input $A = 1$, $B = 1$, $C = 0$, $D = 0$. All the stuck faults for all four inputs are listed, together with the stuck faults for the internal nodes E and F and the output node G. The stuck faults for E and F are only listed once. To distinguish the faulty versions of the circuit from the fault-free version, the gates are labeled according to their output nodes, together with a number. Gate 0 is always the fault-free version. A gate in the fault list is only passed to a gate connected to the output if the faulty value is different from the fault-free value. Thus, $E3$, $E4$, $F1$, and $F2$ appear as inputs to gates in the fault list for G, causing faults $G7$, $G8$, $G9$, and $G10$, respectively. As with parallel fault simulation, it can be seen that for this example, $G1$, $G3$, $G7$, and $G8$, representing $E/0$, $G/0$, $A/0$, and $B/0$, respectively, have different outputs from $G0$ and are therefore detected faults.

To see why concurrent fault simulation is more efficient than parallel fault simulation, suppose that A now changes from 1 to 0. This would cause $E0$, $E2$, and $E4$ to be evaluated. $E1$ and $E3$ would not be evaluated because they both model stuck faults on A. Now, $E0$ is at 0, as are $E2$, $E3$, and $E4$; $E1$ is at 1. The OR gate,

F, and its fault list would not be re-evaluated as neither C nor D change. As faults $E3$ and $E4$ are now the same as $E0$, the corresponding faults in G, $G7$, and $G8$, are removed from the fault list and a fault corresponding to $E1$, say $G11$, is now inserted. Now gate G is evaluated, as E has changed, and faults $G2$, $G3$, $G5$, $G6$, $G9$, $G10$, and $G11$ are evaluated.

It can be seen from Figure 11.8 that, even with this small number of gates, the fault list for G has 10 elements. In practice, the fault lists can be significantly simplified with a little pre-processing of the circuit. It has already been noted that one test can cover a number of faults, and it is possible, in many cases, to deduce that some faults are indistinguishable and that tests for certain faults will always cover certain other faults. The circuit has seven nodes and 14 stuck faults, but it has been shown that only tests for five faults—$A/0$, $C/0$, $D/0$, $A/1$ and $B/1$—are needed and that any other faults are covered by those tests. If this pre-processing is applied, faults $E4$, $F1$, $F2$, $G1$, $G2$, $G3$, $G4$, $G5$, and $G6$ can be eliminated, and $G8$, $G9$, and $G10$ are in turn removed, reducing the fault list for G to one element, $G7$.

Concurrent fault simulation allows efficient selective trace and event scheduling to be used, together with the full range of state and delay models. The major disadvantage is that a significant amount of list processing must be done to propagate faults through the circuit.

The third approach, deductive fault simulation, offers a very similar performance to concurrent fault simulation.

Summary

The principles of digital testing have been introduced. Defects are characterized as logical faults. Test pattern generation algorithms have been described. Parallel and concurrent fault simulation algorithms have also been discussed.

Further Reading

Abramovici, Breuer and Friedman [3] is a very good introduction to fault modeling, test generation, and fault simulation. Also recommended are the books by Wilkins [28] and Miczo [15]. New fault models and algorithms are still being developed, with particular emphasis on delay effects and on sequential systems. *IEEE Design and Test of Computers* provides a quarterly update on developments.

Exercises

11.1 Explain the difference between structural and functional testing.

11.2 What assumptions are made by the SSFM?

11.3 Write down the stuck-at-fault list for the circuit shown in Figure 11.9. Derive tests for A/1 and A/0 and determine which other faults these tests cover. Show that it is not possible to derive a test for G/0.

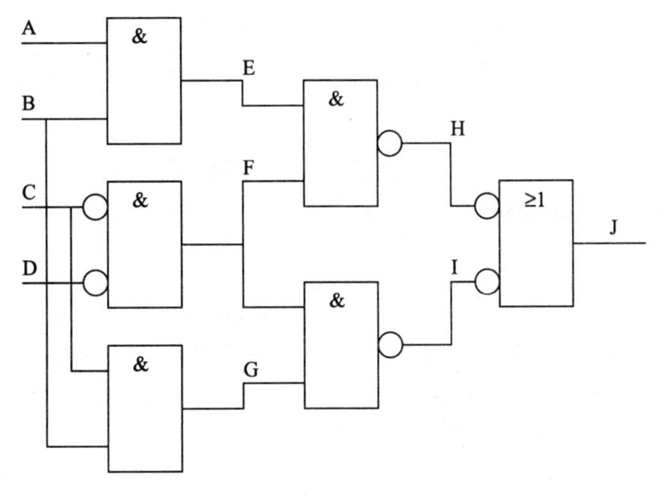

Figure 11.9 Circuit for Exercises 11.3 and 11.4.

11.4 Suggest a test pattern to determine if nodes H and I in Figure 11.9 are bridged together. You should assume that a bridging fault may be modeled as a wired-OR; that is, if either wire is at logic 1, the other wire is also pulled to a logic 1.

11.5 A positive-edge-triggered D-type flip-flop is provided with an active-low asynchronous clear input, and has only its Q output available. By considering the functional behavior of the flip-flop, develop a test sequence for this device for all single-stuck faults on inputs and outputs.

11.6 Describe the four types of crosspoint fault that can occur in a PLA consisting of an AND plane and an OR plane.

11.7 The AND and OR planes of a PLA can be thought of as two NAND planes. What is the minimal set of test patterns required to test an n-input NAND gate?

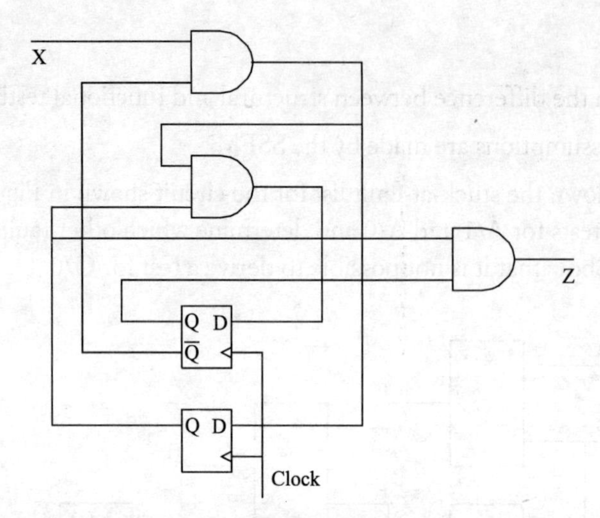

Figure 11.10 Circuit for Exercise 11.8.

11.8 Write down a stuck-fault list for the circuit in Figure 11.10. How, in principle, would a test sequence for this circuit be constructed?

11.9 The circuit shown in Figure 11.11 is an implementation of a state machine with one input and one output. Derive the next state and output equations

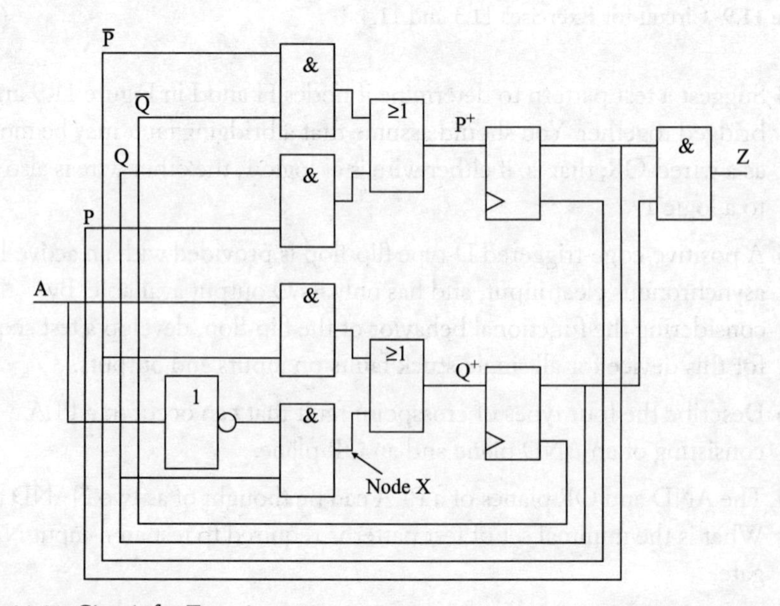

Figure 11.11 Circuit for Exercise 11.9.

and hence show that a parasitic state machine exists, in addition to the intended state machine. Assuming that the initial state of the flip-flops is P = Q = 0, suggest a sequence of input values at A that will cause the output, Z, to have the following values on successive clock cycles: 0110. Hence, show that this sequence of input values can be used to test whether node X is stuck at 0.

11.10 Explain the difference between parallel and concurrent fault simulation.

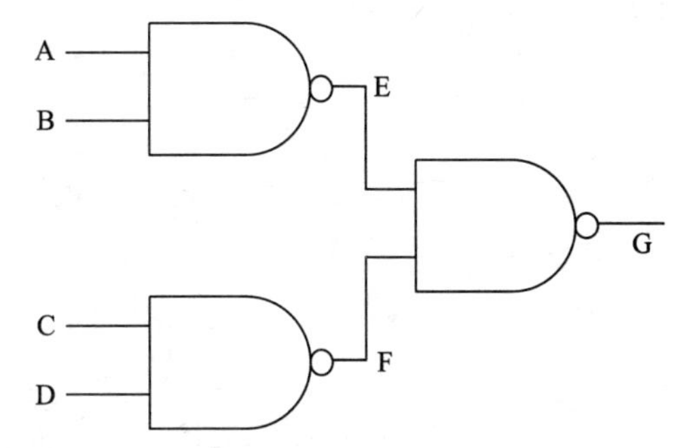

Figure 11.12 Circuit for Exercise 11.11.

11.11 In the circuit of Figure 11.12, A = 1, B = 1, C = 1, D = 0. Derive the fault lists as they would be included in a concurrent fault simulator, assuming that each of the nodes can be stuck at 1 or 0. Show that the fault lists may be significantly simplified if redundant and dominated faults are removed in a pre-processing step.

12 Design for Testability

As noted in the previous chapter, testability for a circuit such as that shown in Figure 12.1 can be expressed in terms of:

- Controllability—the ability to control the logic value of an internal node from a primary input.
- Observability—the ability to observe the logic value of an internal node at a primary output.

The previous chapter discussed methods for finding test patterns for combinational circuits. The testing of sequential circuits is much more difficult because the current state of the circuit as well as its inputs and outputs must be taken into account. Although in many cases it is possible, at least in theory, to derive tests for large complex sequential circuits, in practice it is often easier to modify the design to increase its testability. In other words, extra inputs and outputs are included to increase the controllability and observability of internal nodes.

Testability can be enhanced by ad hoc design guidelines or by a structured design methodology. In this chapter we discuss general ad hoc principles for increasing testability, then look at a structured design technique—the scan path. In the third section, we will see how some of the test equipment itself can be included on an integrated circuit, to provide self-test capabilities. Finally, the scan path principle

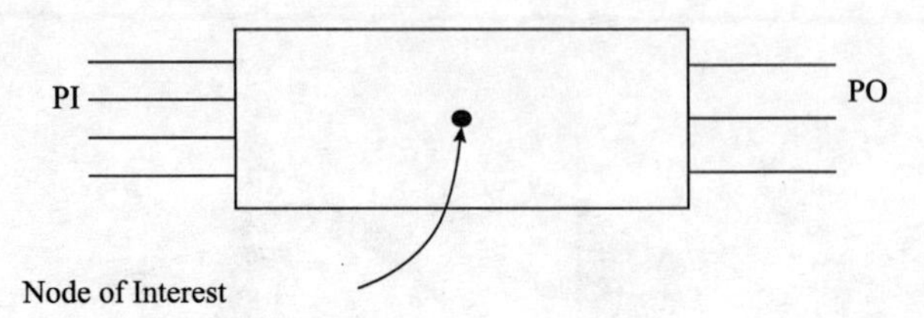

Figure 12.1 Testability of a node.

can be used for internal testing, but it can also be used to test the interconnect between integrated circuits—boundary scan.

12.1 Ad hoc Testability Improvements

If one of the objectives of a design is to enhance the testability of that design, there are a number of styles of design that should be avoided, including the following.

- Redundant logic. As seen in the previous chapter, redundant combinational logic will result in the presence of potentially undetectable faults. This means that the design is not fully testable and also that time may be spent attempting to generate tests for these undetectable faults.
- Asynchronous sequential systems (and, in particular, unstructured asynchronous systems) are difficult to synchronize with a tester. The operation of a synchronous system can be halted with the clock. An asynchronous system is, generally, uncontrollable. If asynchronous design is absolutely necessary, confine it to independent blocks.
- Monostables are sometimes used for generating delays. They are extremely difficult to control and again should be avoided.

On the other hand, there are a number of modifications that could be made to circuits to enhance testability. The single most important of these is the inclusion of some form of initialization. A test sequence for a sequential circuit must start from a known state. Therefore, initialization must be provided for all sequential elements, as shown in Figure 12.2. Any defined state will do—not necessarily all zeros. Multiple initial states can be useful.

The cost of enhancing testability includes that of extra I/O pins (including interfaces, etc), extra components (MUXs), extra wiring, and the degradation of performance because of extra gates in signal paths. In general, there are more things that can go wrong. Against this must be set the benefit that the circuit will be easier

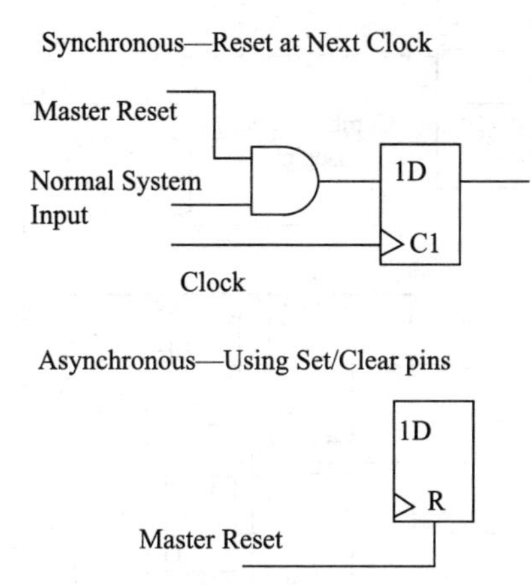

Figure 12.2 Resets add testability.

to test and hence the manufacturer and consumer can be much more confident that working devices are being sold.

12.2 Structured Design for Test

The techniques described in the previous section are all enhancements that can be made to a circuit after it has been designed. A structured design for test method should consider the testability problem from the beginning. Let us restate the problem to see how it can be tackled in a structured manner.

Testing combinational circuits is relatively easy, provided there is no redundancy in the circuit. The number of test vectors is (much) less than $2^{(no.of inputs)}$. Testing sequential circuits is difficult because such circuits have states. A test may require a long sequence of inputs to reach a particular state. Some faults may be untestable because certain states cannot be reached. Synchronous sequential systems, however, can be thought of as combinational logic (next-state and output logic) and sequential logic (registers). Therefore, the key to structured design for test is to separate these two elements.

A synchronous sequential system does not, however, provide direct control of all inputs to the combinational logic, does not allow direct observation of all outputs from the combinational logic, and does not allow direct control or observation of the state variables

The scan-in, scan-out (SISO) principle overcomes these problems by making the state variables directly accessible by connecting all the state registers as a shift register, for test purposes, as shown in Figure 12.3. This shift register has a mode

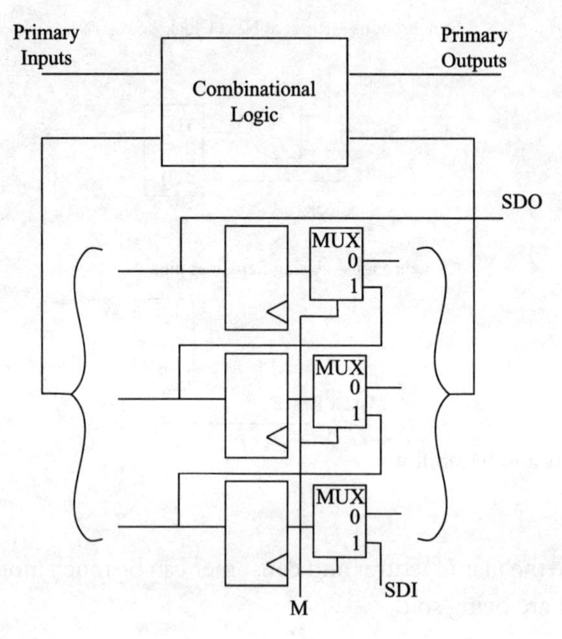

Figure 12.3 SISO principle.

control input, M. In normal, operational mode, M is set to 0. In scan mode, M is set to 1 and the flip-flops form a shift register with the input to the shift register being the scan data in (SDI) pin and the output being the scan data out (SDO) pin.

If the combinational logic has no redundancies, a set of test patterns can be generated for it, as if it were isolated from the state registers. The test patterns and the expected responses then have to be sorted because this test data is applied through the primary inputs and through the state registers, using the scan path. Similarly, the outputs of the combinational logic are observed through the primary outputs and using the scan path.

The scan path is used to test a sequential circuit using the following procedure.

1. Set $M = 1$ and test the flip-flops as a shift register. If a sequence of 1s and 0s is fed into SDI, we would expect the same sequence to emerge from SDO delayed by the number of clock cycles equal to the length of the shift register (n). A useful test sequence would be 00110... which tests all transitions and whether the flip-flops are stable.
2. Test the combinational logic.

   (a) Set $M = 1$ to set the state of the flip-flops after n clock cycles by shifting a pattern in through SDI.

(b) Set $M = 0$. Set up the primary inputs. Collect the values of the primary outputs. Apply one clock cycle to load the state outputs into the flip-flops.
(c) Set $M = 1$ to shift the flip-flop contents to SDO after n–1 clock cycles.

Note that step 2(a) for the next test can be done simultaneously with step 2(c) for the present test. In other words, while the contents of the shift register are being shifted out, new data can be shifted in behind it.

The benefit of using a scan path is that it provides an easy means of making a sequential circuit testable. If there is no redundancy in the combinational logic, the circuit is fully testable. The problem of test pattern generation is reduced to generating tests only for the combinational logic. This can mean that the time to test one device can be greater than would be the case if specific sequential tests had been generated.

The costs of SISO include extra hardware: at least one extra pin for M. SDI and SDO can be shared with other system functions by using multiplexers. An extra multiplexer is needed for each flip-flop and extra wiring is needed for the scan path. Hence, this can lead to performance degradation as the delay through the next state logic is increased. To minimize the wiring, it makes sense to decide the order of registers in the scan path *after* placement of devices on an ASIC or FPGA has been completed. The order of registers is unimportant provided it is known to the tester.

SISO has now become relatively well accepted as a design methodology. Most very large scale integration (VLSI) circuits include some form of scan path, although this is not usually documented.

A number of variations to SISO have been proposed including multiple scan paths—put flip-flops in more than one scan path to shorten the length of each path and to shorten the test time—and partial scan paths whereby some flip-flops are excluded from the scan path.

12.3 Built-In Self-Test

As with all testing matters, the motivation for built-in self-test (BIST or BIT for built-in test) is economic. The inclusion of structures on an integrated circuit or board that not only enhance the testability, but also perform some of the testing, simplifies the test equipment and hence reduces the cost of that equipment. BIST can also simplify test pattern generation because the test vectors are generated internally and allow field testing to be performed, for perhaps years after manufacture. Overall, therefore, BIST should increase user confidence.

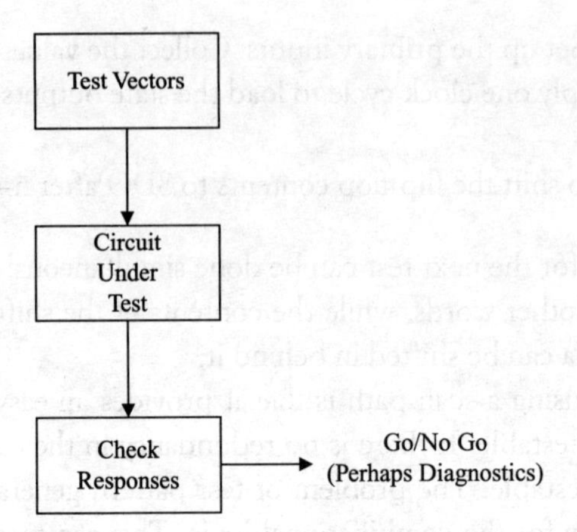

Figure 12.4 BIST principle.

The principle of BIST is shown in Figure 12.4. The test vector generation and checking are built on the same integrated circuit as the circuit under test. Thus, there are two obvious problems: how to generate the test vectors and how to check the responses. It would, in principle, be possible to store pre-generated vectors in ROM. There could, however, be a very large number of vectors. Similarly, it would be possible to have a look-up table of responses.

If an exhaustive test were conducted, all possible test vectors could be generated using a binary counter. This could require a substantial amount of extra combinational logic. A simpler solution is to use an LFSR, introduced in Chapter 5. An LFSR is a pseudo-random number generator that generates all possible states (except the all 0s state) but requires less hardware than a binary counter as shown in Figure 12.5.

A similar structure can be used instead of a look-up table to collect the responses. The single-input signature register (SISR) is shown in Figure 12.6. This is an LFSR with a single data input. The register holds the residue from a modulo-2 division. In other words, it compresses the stream of input data to produce a signature that may be compared, after a certain number of cycles, with a known good signature.

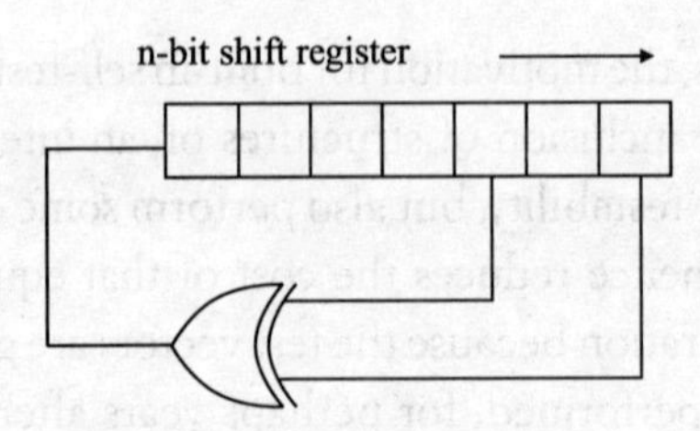

Figure 12.5 LFSR.

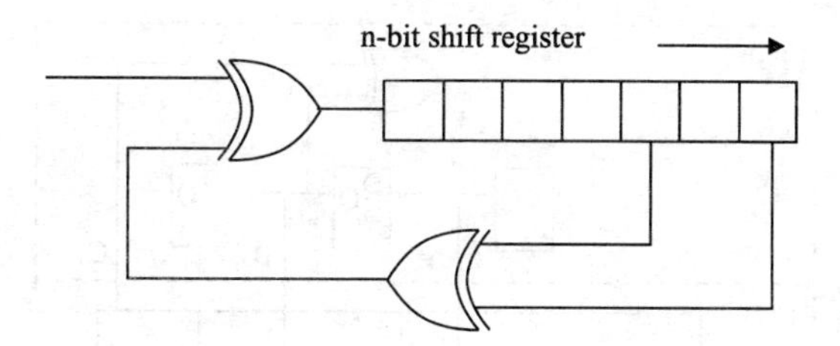

Figure 12.6 SISR.

Another variant is the multiple input signature register (MISR), shown in Figure 12.7. Again, this is a modified LFSR, but with more than one data input. Thus, a number of output vectors can be gathered and compressed. After a number of clock cycles, the signature in the register should be unique. If the circuit contains a fault, the register should contain an incorrect signature, which can easily be checked.

This approach will obviously fail if the MISR is sensitive to errors. The probability that a faulty circuit generates a correct signature tends to 2^{-n} for an n-stage register and long test sequences

12.3.1 Example

For example, consider a circuit consisting of three parts: a three-stage LFSR, a three-stage MISR, and the circuit under test, with the following functions:

$$X = A \oplus B \oplus C$$
$$Y = A \cdot B + A \cdot C + B \cdot C$$
$$Z = \bar{A} \cdot B + \bar{A} \cdot C + B \cdot C$$

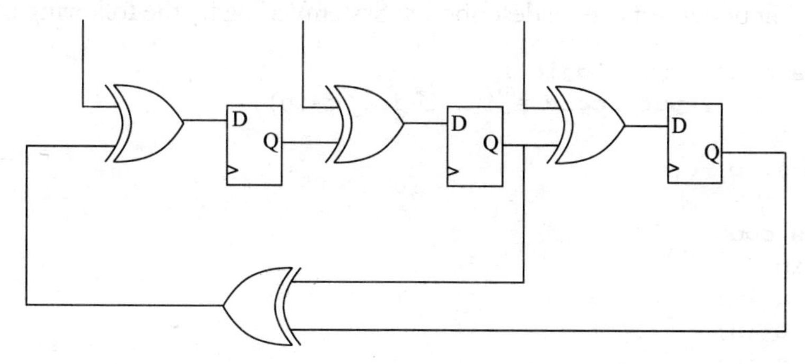

Figure 12.7 MISR.

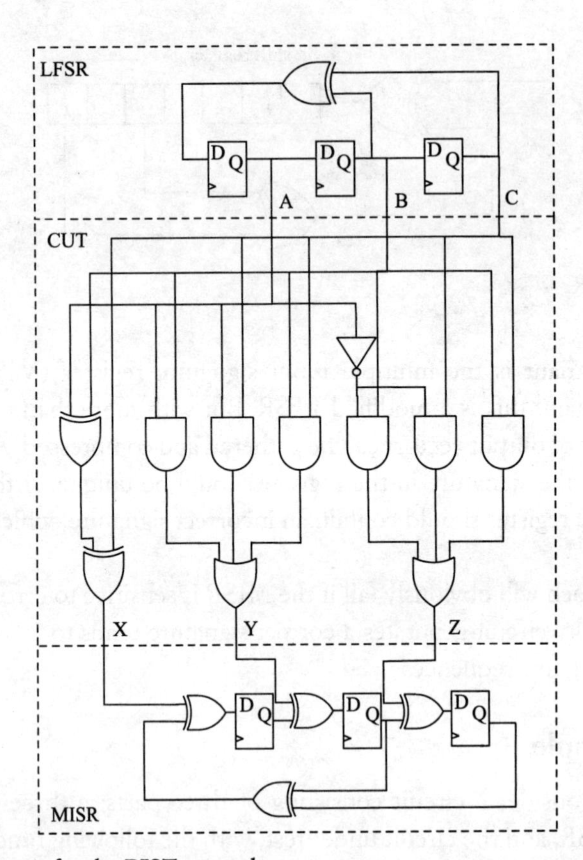

Figure 12.8 Circuit for the BIST example.

The structure of the circuit is shown in Figure 12.8.

In order to see what the correct signature should be, we can perform a simulation. A SystemVerilog model of an LFSR was presented in Chapter 5. This model can easily be adapted to implement an MISR (see the exercises at the end of this chapter). The circuit under test can be described in SystemVerilog by the following model.

```
module cut(output logic x, y, z,
           input logic a_in, b_in, c_in);

logic a, b, c;

always_comb
  begin
  a = a_in;
  b = b_in;
  c = c_in;
  x = a ^ b ^ c;
```

```
  y = (a & b) | (a & c) | (b & c);
  z = (!a & b) | (!a & c) | (b & c);
  end

endmodule
```

The input signals a_in, b_in, and c_in are not used directly because we will insert fault models into those signals later. The test bench for this circuit can, therefore, consist of the following code.

```
module bistex;

  logic clock, n_set;
  logic [2:0] signature, q, z;

  lfsr #(3) l0 (.*);
  misr #(3) m0 (.q(signature), .z(z), .clock(clock),
                .n_set(n_set));
  cut c0 (.a_in(q[2]), .b_in(q[1]), .c_in(q[0]),
          .x(z[2]), .y(z[1]), .z(z[0]));

initial
  begin
        n_set = '1;
  #5ns  n_set = '0;
  #10ns n_set = '1;
  end

always
  begin
  #10ns clock = '0;
  #10ns clock = '1;
  end

endmodule
```

Both the LFSR and MISR are initialized to the 111 state. When the SystemVerilog model is simulated, we get the sequence of states shown in Table 12.1.

The last output of the MISR, 000, is the signature of the fault-free circuit. The intermediate values of the MISR are irrelevant.

We can emulate a stuck fault at the input by changing one of the assignment statements in the CUT. To model a stuck-at 0, the line,

```
a = a_in;
```

is changed to

```
a = '0;
```

Table 12.1 State Sequence for the BIST Example

LFSR Output abc	CUT Output xyz	MISR
111	111	111
011	011	100
001	101	001
100	100	001
010	101	000
101	010	101
110	010	100
111	111	000

(We could, of course, perform a full fault simulation, as described in the previous chapter.) If this perturbed circuit is simulated, the sequence of states is that shown in Table 12.2.

The signature of circuit when a is stuck at 0 is therefore 011. We do not care about the sequence of intermediate states. Hence, a comparison of the value in the MISR with 000 when the LFSR is at 111 would provide a pass/fail test of the circuit. In principle, we could simulate every fault in the circuit and note its signature. This information could be used for fault diagnosis. In practice, of course, we would be assuming that every defect manifests itself as a single stuck fault, so this diagnostic information would have to be used with some caution. Moreover, both the LFSR and MISR could themselves contain faults, which in turn would generate incorrect signatures.

If we run the simulation again for a stuck-at 1, the signature 000 is generated. This is an example of *aliasing*—a fault generates the same signature as the fault-free

Table 12.2 Perturbed State Sequence for the BIST Example

LFSR Output abc	CUT Output xyz	MISR
111	011	111
011	011	000
001	101	011
100	000	100
010	101	010
101	101	000
110	101	101
111	011	011

circuit. The probability of aliasing can be shown to tend to 2^{-n} if a maximal length sequence is used. As there are only three stages to the MISR, the probability of aliasing is 2^{-3} or 1/8. With larger MISRs, the probability of aliasing decreases.

In this example, we have made the LFSR and the MISR the same size and used the complete sequence of inputs once. None of these restrictions is essential. We can use LFSRs of different lengths and we do not need to use all the outputs from the LFSR nor all the inputs to the MISR. We can use a shorter sequence than the complete cycle of the LFSR or we can run through the sequence more than once. In all cases, however, the sequence has to be defined when the circuit is built.

12.3.2 Built-In Logic Block Observation (BILBO)

The LFSR and MISR, described previously, are specialist logic blocks. To include BIST in a circuit using such blocks would require additional registers to those required for normal operation. A scan path reuses the existing registers in a design for testing; in much the same way, built-in logic block observation (BILBO) registers are used both for normal operation and for BIST. A typical BILBO architecture is shown in Figure 12.9. Three control signals are required, which control the circuit as shown in Table 12.3.

To understand the functionality of the circuit, it helps to redraw the functionality of the BILBO when the control signals are set to their different states. Figures 12.10, 12.11, and 12.12 show the normal mode, scan mode, and LFSR/MISR modes, respectively. Note that in the scan, LFSR, and MISR modes, the $\bar{Q}$ output of the flip-flops is used, but inverted before being fed into the next stage. The reset mode

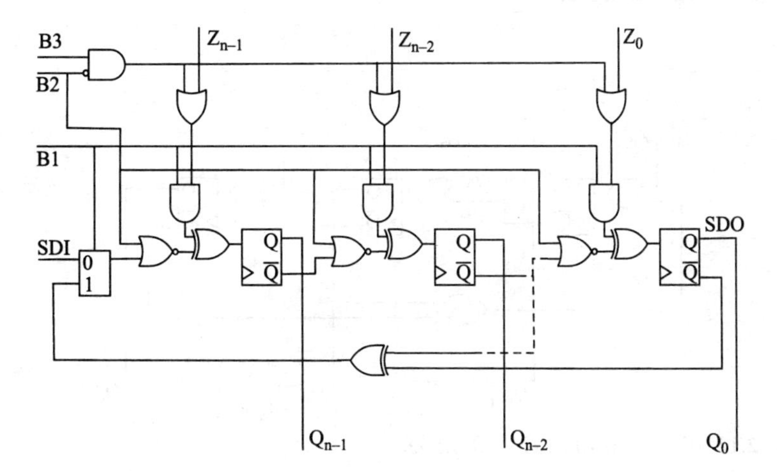

Figure 12.9 BILBO.

Table 12.3 BILBO Modes

B1	B2	B3	Mode
1	1	—	Normal
0	1	—	Reset
1	0	0	Signature analysis MISR
1	0	1	Test pattern generation LFSR
0	0	—	Scan

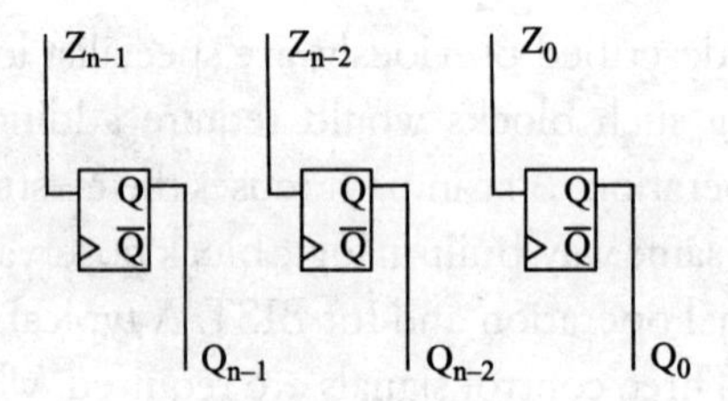

Figure 12.10 BILBO in normal mode.

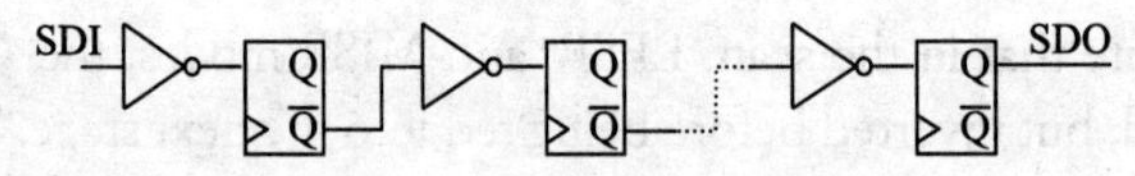

Figure 12.11 BILBO in scan mode.

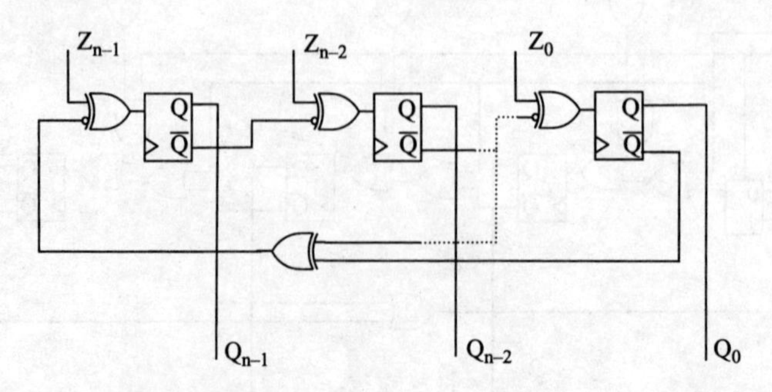

Figure 12.12 BILBO in LFSR/MISR mode.

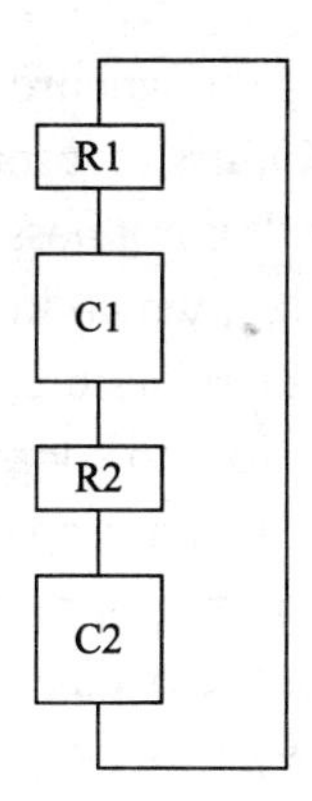

Figure 12.13 Circuit partitioning for a self-test.

synchronously initializes the flip-flops to 0. It was noted in Chapter 5 that an LFSR stays in the all-0s state if it ever enters that state. In LFSR/MISR modes, the BILBO inverts the feedback signal, thus making the all 0s state valid, but there still remain 2^n-1 states in the cycle—one state is excluded from the normal sequence.

Unlike the flip-flops in a scan-path, the flip-flops in a BILBO-oriented system must be grouped into discrete registers. (The scan mode also allows us to link all the BILBOs in a scan-path.) These registers would ideally replace the normal system registers. An example of a system using BILBOs for self-test is shown in Figure 12.13. R1 and R2 are BILBOs, and C1 and C2 are blocks of combinational logic. To test C1, R1 is configured as an LFSR, and R2 is configured as an MISR. Similarly, to test C2, R2 is configured as an LFSR, and R1 is configured as an MISR.

A different arrangement is shown in Figure 12.14. R1, R2, and R3 are BILBOs; C1, C2, and C3 are combinational logic. To test C1, R2 is an LFSR and R1 is an MISR. To test C2, R1 is an LFSR, R2 is an MISR, and so on.

We can therefore use BILBOs to test different structures of combinational logic, but we also need to have some confidence in the correct operation of the BILBOs themselves. Thus, how do we test the BILBOs? The first act in any test must be

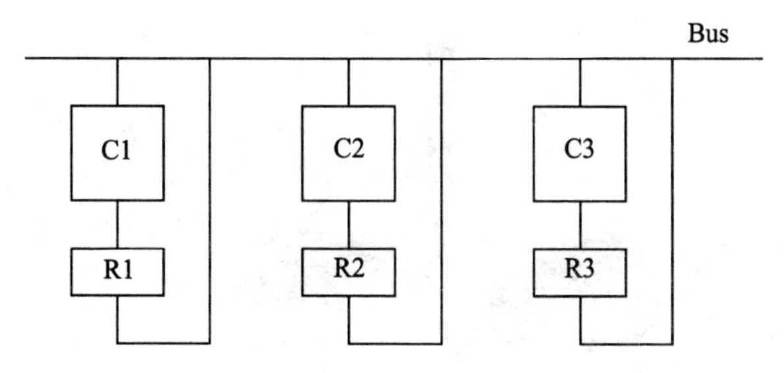

Figure 12.14 Alternate circuit partitioning for a self-test.

initialization. This can be done using the synchronous reset. Then the scan-path can be used to test the flip-flops. This implies that some form of controller is needed to generate the BILBO control signals. It is not possible to test that controller (because a further controller would be needed, which in turn would need to be tested, *ad infinitum*). Therefore, some form of reliable controller is needed to oversee the self-test regime. It makes sense therefore to adopt a "Start Small" strategy, in which part of the system is verified, before being used to test a further part of the system. If the system includes some form of microprocessor, software-based tests can be performed once the microprocessor has been checked.

Before adopting BIST in a design, the cost and effectiveness of the strategy must be considered. There is, of course, the cost of additional hardware—just over four gates per flip-flop for a BILBO-based design, together with the cost of a test controller and the additional assorted wiring. This means that there will be an increased manufacturing cost. The extra hardware means that the reliability of the system will be decreased—there is more to go wrong. There is also likely to be some performance degradation as the hardware between flip-flops is increased. The incorporation of BIST means the complexity of the design and hence the time taken to do the design is increased. On the other hand, using BIST means that the costs of test pattern generation disappear and that the equipment needed to test integrated circuits can be simplified. Moreover, the tests can be performed every time the circuit is switched on, not merely once at the time of manufacture.

12.4 Boundary Scan (IEEE 1149.1)

The techniques described thus far in this chapter have been oriented toward integrated circuits, in which controllability and observability may be limited. Circuits built from discrete gates on printed circuit boards (PCBs) are generally considered easier to test because it is possible to gain access to all the nodes of the circuit using a probe, as shown in Figure 12.15, or a number of probes arranged as a "bed-of-nails." This assumption has become invalid in recent years for the following reasons.

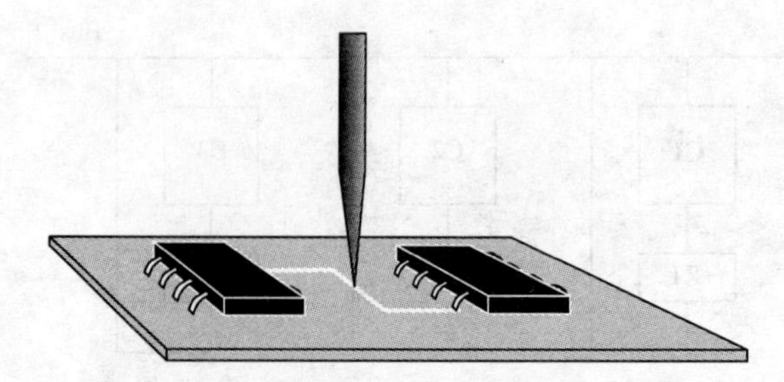

Figure 12.15 Probe testing.

- It is not possible to test mounted ICs (the pins may be connected together).
- PCBs now often have more than 20 layers of metal, so deep layers cannot be reached.
- The density of components on a PCB is increasing. Multi-chip modules (MCMs) take the chip/board concept further and have unpackaged integrated circuits mounted directly on a silicon substrate.

Boundary scan is a technique for testing the interconnect on PCBs and for testing ICs mounted on PCBs. As before, both the ICs and the empty PCB can be tested, but boundary scan replaces the step of testing the loaded PCB with a "bed-of-nails" tester. The bed-of-nails approach has also been criticized because of "back-driving"—in order to test a single gate, its inputs would be forced to particular logic values, which also forces those logic values onto the outputs of other gates. This is not how gates are designed to work and may cause them damage.

The principle of boundary scan is to allow the *outputs* of each IC to be controlled and the *inputs* to be observed. For example, consider the faults shown in Figure 12.16. These faults are external to the integrated circuits and have arisen as a result of assembling (fault-free) ICs onto a PCB. Instead of using mechanical probes to access the board, the faults are sensitized electrically. The outputs of the left-hand ICs in Figure 12.16 are used to establish test patterns, and the inputs of the right-hand IC are used to observe the responses. Therefore, we need to control and observe the output and input pins, respectively, of the integrated circuits. This can be done by connecting those pins, on the *boundary* of the integrated circuits, into a *scan* path, using special logic blocks at each input and output.

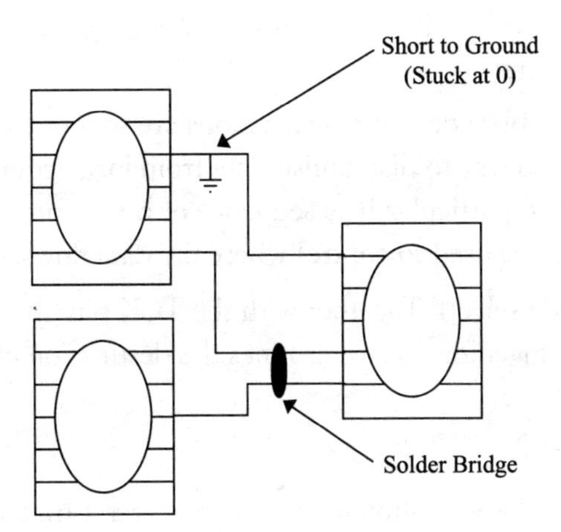

Figure 12.16 Circuit board faults.

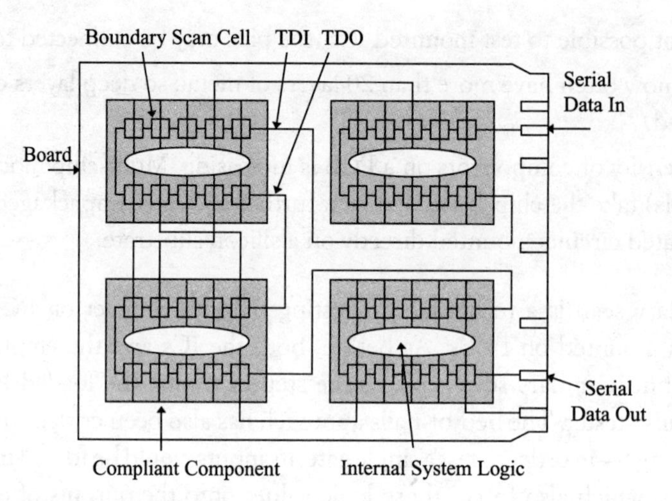

Figure 12.17 Board with boundary scan.

Figure 12.17 shows how the input and output pins of all the ICs on a board are connected together in a scan path. Each IC has dedicated pins to allow the scan path to pass through it. These pins are labeled as TDI (test data in) and TDO (test data out). In addition, control pins will be needed. The various ICs on a board may come from different manufacturers. For boundary scan to work, the ICs need to use the same protocols. Therefore, an IEEE standard, 1149.1, has been defined. This standard arose from the work of the Joint Test Action Group (JTAG). The term JTAG is therefore often used in reference to the boundary scan architecture.

Every boundary scan compliant component has a common test architecture, shown in Figure 12.18. The elements of this architecture are as follows.

1. Test access port (TAP)—The TAP consists of four or five additional pins for testing. The pins are:
   - TDI and TDO. Both data and instructions are sent to ICs through the scan path. There is no way to distinguish data from instructions, or indeed to determine which particular IC a sequence of bits is intended to reach. Therefore, TMS is used to control where the data flows.
   - TMS (test mode select). Together with the TCK pin, the TMS pin is used to control a state machine that determines the destination of each bit arriving through TDI.
   - TCK (test clock)
   - TRST (test reset) is an optional asynchronous reset (not shown in Figure 12.18).

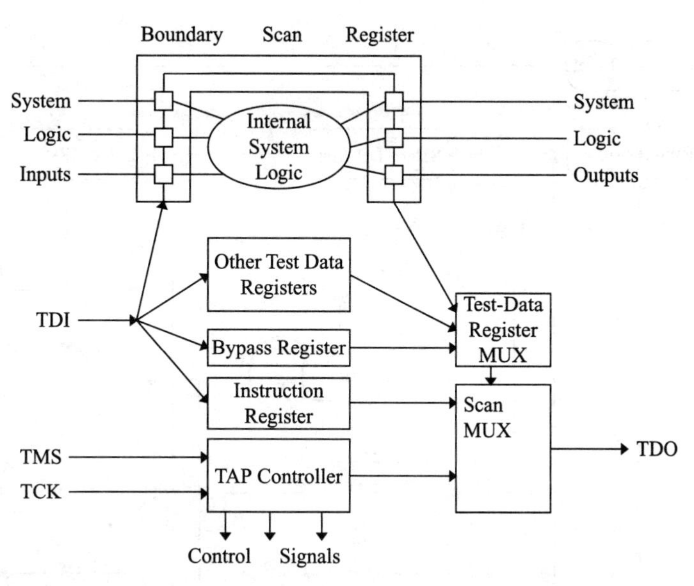

Figure 12.18 Boundary scan architecture.

2. TAP controller—This is a 16-state machine that controls the test. The inputs to the state machine are TCK and TMS. The outputs are control signals for other registers. The state chart of the TAP controller is shown in Figure 12.19. Notice that a sequence of five 1s on TMS in successive clock cycles will put the state machine into the test-logic-reset state from any other state. The control signals derived from the TAP controller are used to enable other registers in a device. Thus, a sequence of bits arriving at TDI can be sent to the instruction register or to a specific data register, as appropriate.
3. Test data registers—A boundary scan compliant component must have all its inputs and outputs connected into a scan path. Special cells, described later, are used to implement the scan register. In addition, there must be a bypass register of 1 bit. This allows the scan path to be shortened by avoiding the boundary scan register of a component. Other registers may also be included, for example, an IC might include an identification register, the contents of which could be scanned out to ensure that the correct device had been included on a PCB. Similarly, the internal scan path of a device could be made accessible through the boundary scan interface. Some programmable logic manufacturers allow the boundary scan interface to be used for programming devices. Thus, the configuration register is another possible data register.

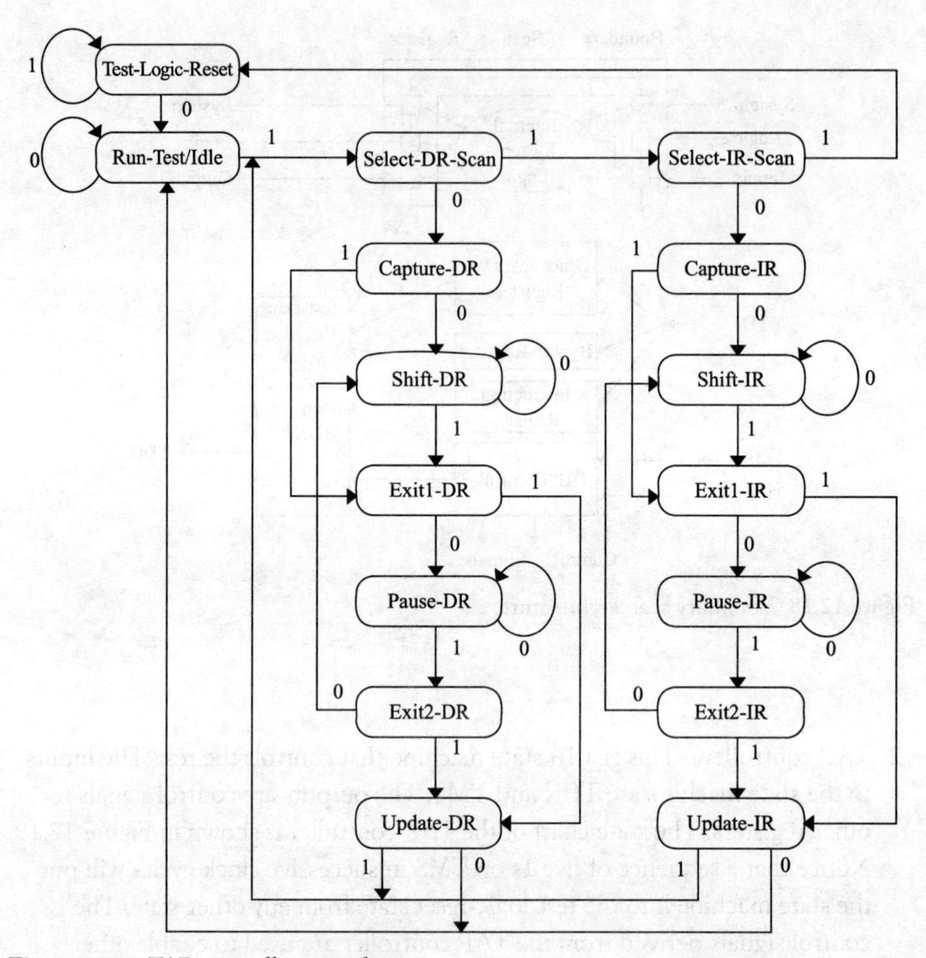

Figure 12.19 TAP controller state diagram.

4. Instruction register—This register has at least 2 bits, depending on the number of tests implemented. It defines the use of test data registers. Further control signals are derived from the instruction register.

The core logic is the normal combinational and sequential logic of the device. This core logic may (should) contain a scan path and may also contain BIST structures.

A typical boundary scan cell is shown in Figure 12.20. This cell can be used for an input or an output pin. For an input pin, IN is connected to the pin, OUT is connected to the device core; for an output pin, IN comes from the core, OUT goes to the pin. Other designs of boundary scan cell are possible.

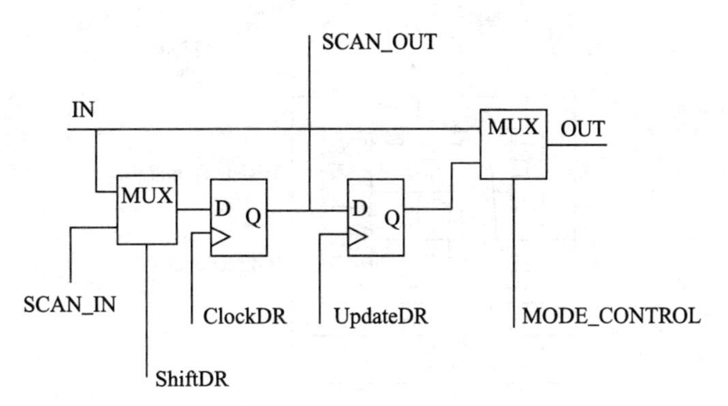

Figure 12.20 Boundary scan cell.

The boundary scan cell has four modes of operation.

1. Normal mode. Normal system data flows from IN to OUT.
2. Scan mode. ShiftDR selects the SCAN_IN input, and ClockDR clocks the scan path. ShiftDR is derived from the similarly named state in the TAP controller of Figure 12.19. ClockDR is asserted when the TAP controller is in state Capture-DR or Shift-DR. (Hence, of course, the boundary scan architecture is not truly synchronous!)
3. Capture mode. ShiftDR selects the IN input and data is clocked into the scan path register with ClockDR to take a snapshot of the system.
4. Update mode. After a capture or scan, data from the left flip-flop is sent to OUT by applying one clock edge to UpdateDR. Again, this clock signal comes from the TAP controller when it is in state Update-DR. The TAP controller then enters the run test state and MODE_CONTROL is set as appropriate according to the instruction held in the instruction register (see the following).

For normal input and output pins, the boundary scan cells are the only logic between the core and the IC pins. The only cases where logic is permitted between the boundary scan cell and an external pin are shown in Figure 12.21.

A number of instructions may be loaded into the instruction register. These allow specific tests to be performed. During test execution, the TAP controller is in the run test state. Three of these tests are mandatory; the remaining tests are optional. Some of these tests are as follows.

- EXTEST (Mandatory). This instruction performs a test of the system, external to the core logic of particular devices. Data is sent from the output

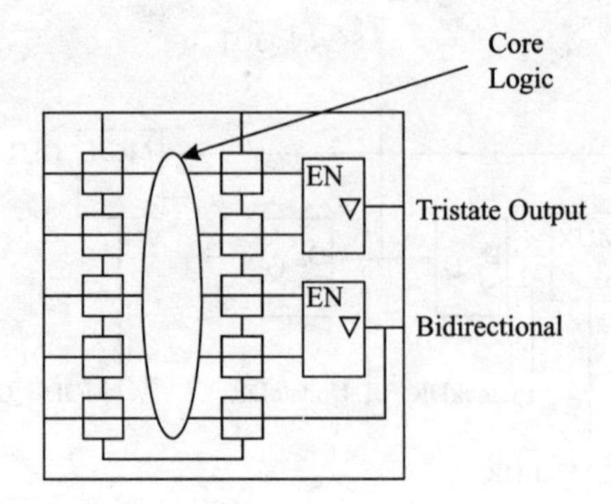

Figure 12.21 Logic outside boundary scan path.

boundary scan cells of one device, through the pads and pins of that device, along the interconnect wiring, through the pins and pads of a second device, and into the input boundary scan cells of that second device. Hence, a complete test of the interconnect from one IC core to another is performed.

- SAMPLE/PRELOAD (Mandatory). This instruction is executed before and after the EXTEST and INTEST instructions to set up pin outputs and to capture pin inputs.
- BYPASS (Mandatory). This instruction selects the bypass register to shorten the scan path.
- RUNBIST (Optional). Runs a built-in self-test on a component.
- INTEST (Optional). This instruction uses the boundary scan register to test the internal circuitry of an IC. Although such a test would normally be performed before a component is mounted on a PCB, it might be desirable to check that the process of soldering the component onto the board has not damaged it. Note that the internal logic is disconnected from the pins, so if pins have been connected together on the board, that will have no effect on the standard test.
- IDCODE, USERCODE (Optional). These instructions return the identification of the device (and the user identification for a programmable logic device). The code is put into the scan path.
- CONFIGURE (Optional). An SRAM-based FPGA needs to be configured each time power is applied. The configuration of the FPGA is held in registers. These registers can be linked to the TAP interface. This clearly saves pins as the configuration and test interfaces are shared.

The MODE_CONTROL signal of Figure 12.20 is set to select the flip-flop output when instructions EXTEST, INTEST, and RUNBIST are loaded in the instruction register. Otherwise, the IN input is selected.

Testing a board with boundary scan components is in many ways similar to testing a component with a scan path. First, the boundary scan circuitry itself must be tested for faults such as a broken scan path or a TAP failure. Then, interconnect and other tests can be performed. The boundary scan path allows nodes to be controlled from one point in the scan path and observed at another point. Test patterns for the interconnect (and for non-boundary scan compliant components) have to be derived in much the same way that tests for logic are determined. These tests and the appropriate instructions have to be loaded into the registers of boundary scan components in the correct order. This process is clearly complex to set up and really has to be automated.

An example of how boundary scan might be included on an IC is shown in Figure 12.22. The basic circuit has two D-type flip-flops with a clock and reset. The D, Q, clock, and reset pins have boundary scan cells included as shown in the figure. A TAP controller and instruction and bypass registers are included, together with the four extra pins.

The costs of implementing boundary scan on an IC include the cost of a boundary scan cell for each pin, the TAP controller, the 1-bit bypass register, the instruction register, and four extra pins. There will be extra wiring on the PCB.

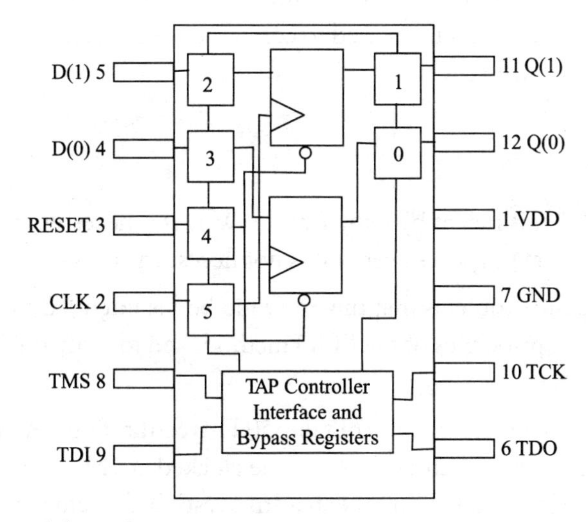

Figure 12.22 IC with boundary scan.

On the other hand, there can be significant benefits. The fault coverage of a PCB can be close to 100%. Boundary scan is easy to implement on a PCB requiring four pins on an edge connector. Specialist, expensive test equipment, such as a bed-of-nails tester, is not needed. Indeed, it is possible to implement a boundary scan tester using little more than a standard PC or workstation. Tests can be performed on ICs after they have been mounted on the PCB, so field testing is easy. Because the test circuitry is independent of normal system functions, it is possible to monitor the inputs and outputs of ICs in normal operation, thus providing debugging functions.

Summary

The testability of a circuit can be improved by modifying the circuit design. The simplest modifications include providing asynchronous resets to every register and avoiding redundant and other uncontrollable logic. SISO separates the sequential from the combinational logic, reducing test generation to a purely combinational circuit problem. BIST can reduce manufacturing costs by putting much of the test circuitry on the chip. Boundary scan uses the SISO principle to allow complex PCBs to be tested. These various techniques can be combined.

Further Reading

The books by Abramovici, Breuer and Friedman [3], Miczo [15], and Wilkins [28] all describe design for test methods. Boundary scan is now incorporated into many FPGAs, and the TAP interface is used to configure the internal logic. Details are on the manufacturers' Web sites.

Exercises

12.1 Explain what is meant by initialization. Why is it necessary to initialize a circuit for test purposes even if it is not necessary in its system function?

12.2 What are the problems that the SISO method is intended to overcome? Explain the principles of the SISO method, and identify the benefits and costs involved.

12.3 A certain integrated circuit contains 50 D-type flip-flops. Assuming that all states are reachable, and that it may be clocked at 1MHz, what is the minimum time needed for an exhaustive test? If the same integrated circuit

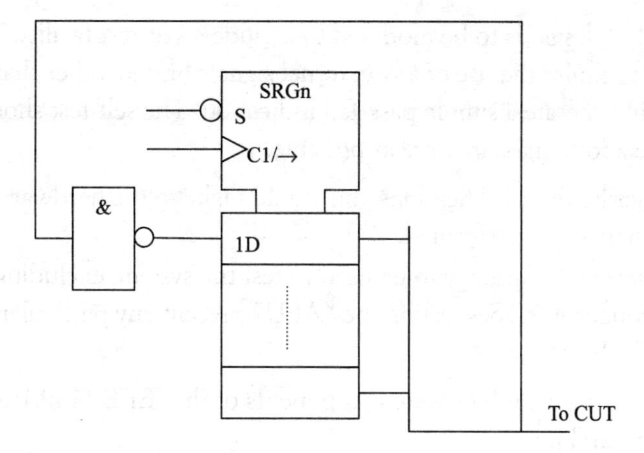

Figure 12.23 Circuit for Exercise 12.4.

is designed with a full scan-path and if all the combinational logic may be fully tested with 200 test vectors, estimate the time now required to complete a full test.

12.4 Show that the circuit of Figure 12.23 is a suitable test generator for an n-input NAND gate. Hence, suggest a suitable BIST structure for each of the NAND planes in a PLA.

12.5 Figure 12.24 shows the structure of a simple CPU (reproduced from Chapter 7). There is a single bus, 8 bits wide. "PC," "IR," "ACC," "MDR," and "MAR" are 8-bit registers. "Sequencer" is a state machine with inputs from the "IR" block and from other points in the system and with outputs that control the operation of the "ALU" and that determine which register drives the bus.

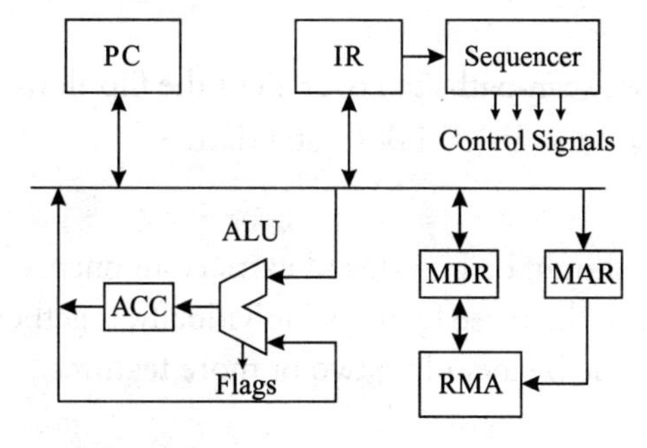

Figure 12.24 CPU datapath for Exercise 12.5.

The CPU design is to be modified to include a self-test facility. This self-test will not require the use of any external signals or data other than the clock and will generate a simple pass/fail indication. The self-test should require as little additional hardware as possible.

(a) Describe the modifications you would make to the hardware to allow a self-test to be performed.

(b) Describe the strategy to be used to test the system, excluding the "Sequencer." Does testing the "ALU" present any particular difficulties?

12.6 What are the main hardware components of the IEEE 1149.1 boundary scan test architecture?

12.7 Figure 12.19 shows the state transition diagram of the boundary scan TAP controller. Assuming that the instruction for an EXTEST is 10 for a particular IC, what sequence of inputs needs to be applied to the TAP of that IC to load the pattern 1010 into the first four stages of the boundary scan register of the IC and to run an EXTEST? (Note that the least significant bits should be loaded first.)

12.8 If the outputs of four boundary scan register stages are connected to the inputs of four similar register stages in a second IC, show, in principle, how the test sequence from Exercise 12.7 can be extended to capture the responses of the interconnect. What assumptions have you made about the connection of the test structures on the two ICs?

12.9 A particular integrated circuit has 2000 flip-flops and 5000 other gates. The package has 52 pins, including power, ground, clock, and reset. All the buses are 16 bits wide. A new version of the circuit is to be built. Before redesigning the circuit, the manufacturer would like an estimate of the costs of:

(a) One or more scan-paths to cover all of the flip-flops

(b) Boundary scan to IEEE 1149.1 standard

(c) BIST

The estimates should be in terms of extra components and pins and should consider each of the three features individually, together with any savings that may be made by including two or more features.

Table 12.4 TAP Controller Outputs for Exercise 12.11

Signal	State(s)
UpdateDR	Update-DR
ClockDR	Capture-DR Shift-DR
ShiftDR	Shift-DR
UpdateIR	Update-IR
ClockIR	Capture-IR Shift-IR
ShiftIR	Shift-IR

12.10 Modify the SystemVerilog model of the LFSR from Chapter 5 to implement an n-stage MISR. Hence, write a model of an n-bit BILBO register.

12.11 Write a synthesizable SystemVerilog model of the IEEE 119.1 TAP controller. The outputs shown in Table 12.4 should be asserted.

13 Asynchronous Sequential Design

The sequential circuits described in Chapters 5, 6, and 7 are synchronous. A clock is used to ensure that all operations occur at the same instant. This avoids the problems of hazards because such transient effects can be assumed to have died away before the next clock edge. Therefore, irredundant logic can be used, which then makes the combinational parts of the circuits fully testable, at least in theory. The flip-flops used in synchronous design are, however, asynchronous internally. In this chapter, we consider the design of asynchronous elements, and use a SystemVerilog simulator to illustrate the difficulties of asynchronous design.

13.1 Asynchronous Circuits

Throughout this book, the emphasis has been on the design of *synchronous* sequential circuits. State information or other data has been loaded into flip-flops at a clock edge. Asynchronous inputs to flip-flops have been used, but *only* for initialization. A common mistake in digital design is to use these asynchronous inputs for purposes other than initialization. This mistake is made either because of inexperience or because of a desire to simplify the logic in some way. Almost inevitably, however, circuits designed in such a manner will cause problems by malfunctioning or because subsequent modification or transfer to a new technology will cause the assumptions made in the design to become invalid.

Synchronous sequential design is almost overwhelmingly preferred and practiced because it is easier to get right than asynchronous design. Simply connecting logic to the asynchronous inputs of flip-flops is almost always wrong. Structured design techniques exist for asynchronous design and this chapter describes the design process and its pitfalls. It should be noted, however, that we are primarily concerned with the design of circuits comprising a few gates. It is possible to design entirely asynchronous systems, but such methodologies are still the subject of research. Nevertheless, as clock speeds increase, some of the complex timing issues described here will become relevant. It is increasingly difficult to ensure that a clock edge arrives at every flip-flop in a system at *exactly* the same instance. Systems may consist of synchronous islands that communicate asynchronously. To ensure such communications are as reliable as possible, specialized interface circuits will need to be designed, using the techniques described in this chapter.

Although, as noted above, this book has been concerned with synchronous systems, reference was made to the synthesis of asynchronous elements (latches) in Chapter 10. At present, synthesis tools are intended for the design of synchronous systems, normally with a single clock. This is particularly true of synthesis tools intended for FPGA design. The SystemVerilog construct

```
assign q = c ? d : q;
```

would be synthesized to an asynchronous sequential circuit structure. Similarly, the sequential block

```
always_latch
  if (c)
    q <= d;
```

would also be synthesized to an asynchronous latch. In both cases, q explicitly holds onto its value unless c is asserted. It might be thought that the circuit structures created by a synthesis tool for the two cases would be identical. In general, this is not so. The first case is exactly the same as writing

```
assign q = (d & c) | (q & ~c);
```

Hence, a synthesis tool would create an inverter, two AND gates, and an OR gate (or an inverter and three NAND gates). On the other hand, a compliant synthesis tool would infer the existence of a latch from the incomplete **always_latch** statement of the second case, and use a latch from a library (while also issuing a warning message, in case the incomplete **if** statement were a coding error). The latch created by Boolean minimization and the library latch are not the same. Indeed, the Verilog RTL synthesis standard IEEE 1364.1 explicitly forbids the use of concurrent

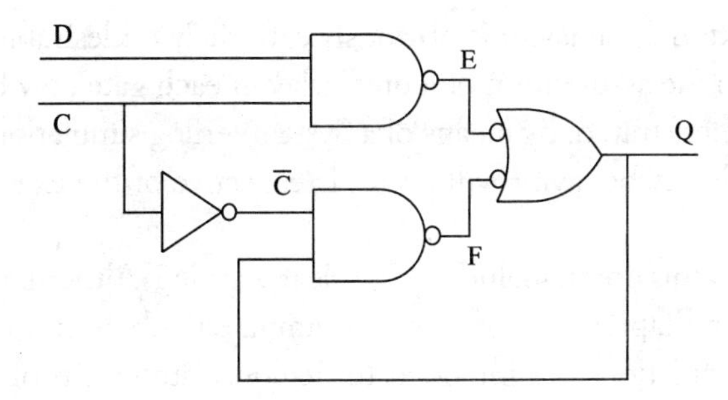

Figure 13.1 Basic D latch.

assignments of the form shown, while permitting the use of incomplete **if** and **case** statements.

To see why, assume that the circuit has been implemented directly, as shown in Figure 13.1. This circuit should be compared with that of Figure 2.13. Indeed, the following analysis is comparable with that of Section 2.4. Let us assume that each gate, including the inverter, has a delay of 1 unit of time, for example, 1 ns. Initially, Q, D, and C are at logic 1. C then changes to 0. From the analysis of Section 2.4, we know that this circuit contains a potential hazard. When we draw a timing diagram for this circuit, as shown in Figure 13.2, this hazard appears at Q. This hazard is propagated back to F, which causes Q to change *ad infinitum*. Hence, the circuit oscillates. The causality between F and Q is not shown in Figure 13.2

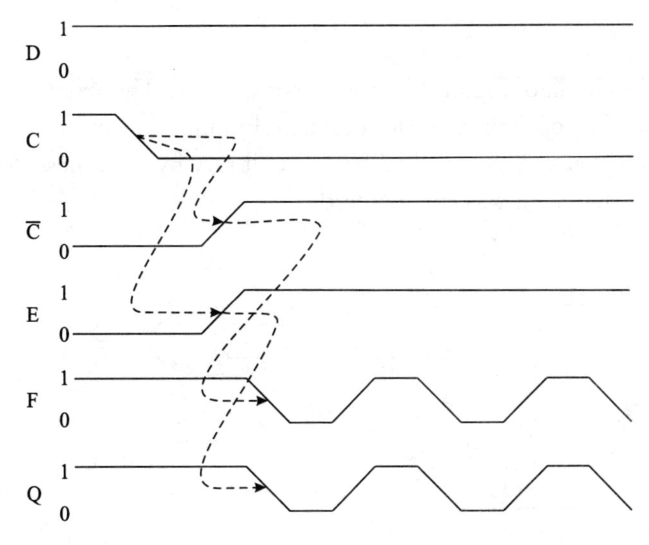

Figure 13.2 Timing diagram for the circuit of Figure 13.1.

for clarity. This kind of behavior is obviously extremely undesirable in a sequential circuit. Although the assumption of a unit delay in each gate may be unrealistic, it can easily be demonstrated, by means of a SystemVerilog simulation, that a hazard, and hence, oscillatory behavior will occur, irrespective of the exact delays in each gate.

We should, at this point, include a very clear warning. Although we will use SystemVerilog in this chapter to model and to simulate the behavior of asynchronous circuits, these simulations are intended to demonstrate that problems *may* exist. It is extremely difficult to accurately predict, by simulation, *exactly* how a circuit will behave, particularly when illegal combinations of inputs are applied. The spurious effects result from voltage and current changes within electronic devices, not transitions between logic values.

The solution to the problem of oscillatory behavior is, as stated in Section 2.4, to include redundant logic by way of an additional gate. Thus,

$$Q^+ = D \cdot C + Q \cdot \bar{C} + D \cdot Q$$

or

$$Q^+ = \overline{\overline{D \cdot C} \cdot \overline{Q \cdot \bar{C}} \cdot \overline{D \cdot Q}}$$

where Q^+ represents the "next" value of Q. The redundant gate, $\overline{D \cdot Q}$, has a 0 output while D is 1. Therefore, Q is held at 1.

The expression for Q^+ can be rearranged:

$$Q^+ = D \cdot C + Q \cdot (\bar{C} + D)$$

Hence, the circuit of Figure 13.3 can be constructed. This would not and could not be generated by optimizing logic equations, but instead would exist in a library. It is this circuit that would be called from the library by a synthesis tool when an incomplete **if** statement was encountered.

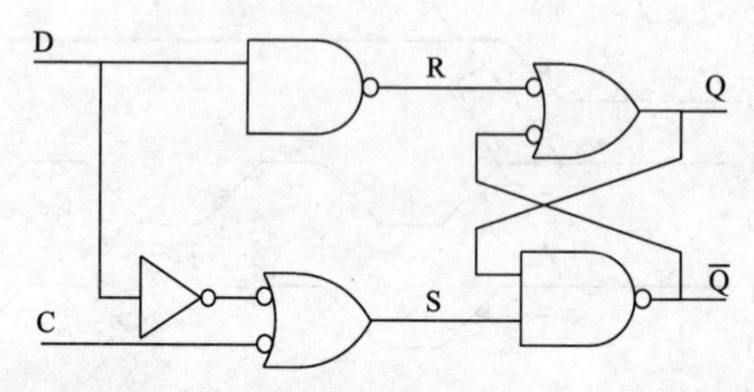

Figure 13.3 D latch with hazard removed.

Table 13.1 Truth Table for an RS Latch

R	S	Q^+	$\bar{Q}^+$
0	0	1	1
0	1	1	0
1	0	0	1
1	1	Q	$\bar{Q}$

13.2 Analysis of Asynchronous Circuits

13.2.1 Informal Analysis

The operation of the D latch of Figure 13.3 is relatively straightforward. The key is the operation of the cross-coupled NAND gates. Two NAND (or NOR) gates connected in this way form an RS latch with the truth table given in Table 13.1. (An RS latch built from NOR gates has a similar truth table, but with the polarities of R and S reversed.)

The input $R = S = 0$ is normally considered illegal because it forces the outputs to be the same, contradicting the expected behavior of a latch.

The D latch of Figure 13.3 contains an RS latch, in which R and S are controlled by two further NAND gates. When C is at logic 0, R and S are at 1. Therefore, the latch holds whatever value was previously written to it. When C is 1, S takes the value of D and R takes the value of $\bar{D}$. From the truth table, we can see that Q therefore takes the value of D. We can further note that the signal paths from D to the outputs are unequal because of the inverter. It is therefore reasonable to assume that if D and C were to change at the same time, the behavior of the latch would be unpredictable.

Figure 13.4 shows the circuit of a positive edge-triggered D flip-flop. We will attempt to analyze this circuit informally, but this analysis is intended to show that a formal method is needed. Let us first deal with the "asynchronous" set and reset.[1] If S is 0 and R is 1, Q is forced to 1 and $\bar{Q}$ is forced to 0, according to the truth table. Similarly, if S is 1 and R is 0, Q is forced to 0 and $\bar{Q}$ is forced to 1. Under normal synchronous operation, S and R are both held at 1, and therefore can be ignored in the following analysis. Note, however, that if both S and R are held at 0, both Q and $\bar{Q}$ go to 1. Hence, this condition is usually deemed to be illegal.

1. At this level, all of the inputs are asynchronous, of course. Synchronous design works because we follow certain conventions about the use of inputs, not because particular inputs are special.

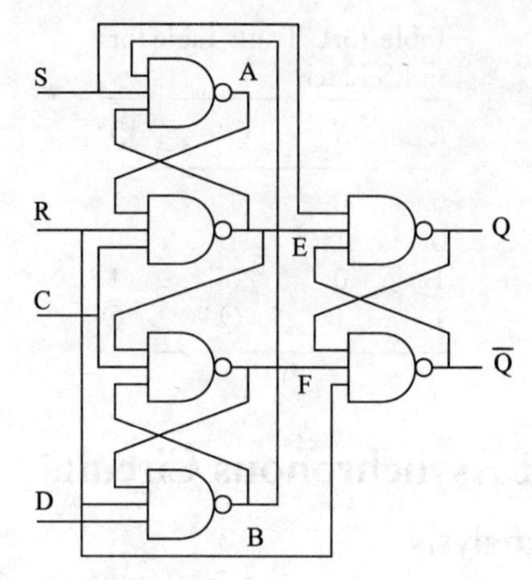

Figure 13.4 Positive edge-triggered D flip-flop.

Let us consider the effects of changes at the D and C inputs, while $R = S = 1$. If C is at 0, then both E and F are at 1 and therefore Q and $\bar{Q}$ are held. If D is at 0, internal nodes A and B are at 0 and 1, respectively. If D is at 1, A is 1 and B is 0. Therefore, D can change while the clock is low, causing A and B to change, but further changes to E and F are blocked by the clock being low.

When the clock changes from 0 to 1, *either* D is 0, and hence A is 0 and B is 1, which force E to 1 and F to 0 and therefore, Q to 0 and $\bar{Q}$ to 1, *or* D is 1, A is 1, B is 0 and therefore E is 0, F is 1, Q is 1, and $\bar{Q}$ is 0. Therefore, when the clock changes, it is assumed that A and B are stable. Hence, there is a *setup time* in which any change in D must have propagated to A before the clock edge.

While the clock is 1, D can again change without affecting the outputs. Two conditions are possible: (a) D was 0 at the clock edge, and hence A is 0, B is 1, E is 1, and F is 0. If D changes to 1, there will be no change to B because F is 0 and hence B is always 1 or (b) D was 1 at the clock edge, thus A is 1, B is 0, E is 0, and F is 1. If D changes to 0, B changes from 0 to 1, but as E is 0, this change is not propagated to A. Therefore, again, the output is unaffected. The falling clock edge forces both E and F to 1 again.

It is apparent that this descriptive, intuitive form of analysis is not sufficient to adequately describe the behavior of even relatively small asynchronous circuits. Moreover, it would be impossible to design circuits in such a manner. It is possible to use a SystemVerilog simulator to verify the behavior of such circuits, but we need a formal analysis technique.

13.2.2 Formal Analysis

Before proceeding with the formal analysis of both the D latch and the edge-triggered D flip-flop, we need to state a basic assumption. The *fundamental mode* restriction states that only one input to an asynchronous circuit may change at a time. The effects of an input change must have propagated through the circuit, and the circuit must be stable before another input change can occur. The need for this restriction can be seen from the two circuits already considered. If D changes at almost the same time as the clock, unequal delay paths mean that internal nodes are not at expected, consistent values, and unpredictable behavior may result. In the worst case, the output of a latch or flip-flop may be in an intermediate, *metastable* state, that is neither 0 nor 1. We will return to metastability later.

In order to perform a formal analysis, we have to break any feedback loops in the circuit. Of course, we don't actually change the circuit, but for the purposes of the analysis, we pretend that all the gate delays in the circuit are concentrated in one or more *virtual buffers* in the feedback loops. The gates are therefore assumed to have zero delays. The D latch is redrawn in Figure 13.5. Note that there is only one feedback loop in this circuit, although at first glance the cross-coupled NAND gate pair might appear to have two feedback loops. If the one feedback loop were really broken, the circuit would be purely combinational, which is sufficient. In Figure 13.5, the input to the virtual buffer is labeled as Y^+, while the output is labeled as Y. Y is the *state variable* of the system. This is analogous to the state variable in a synchronous system. Y^+ is the next state. The system is *stable* when Y^+ is equal to Y. In reality, of course, Y^+ and Y are two ends of a piece of wire and must have the same value, but, to repeat, for the purpose of analysis, we pretend that they are separated by a buffer having the aggregate delay of the system. Note that we separate the state variable from the output, although in this case, Q and Y^+ are identical.

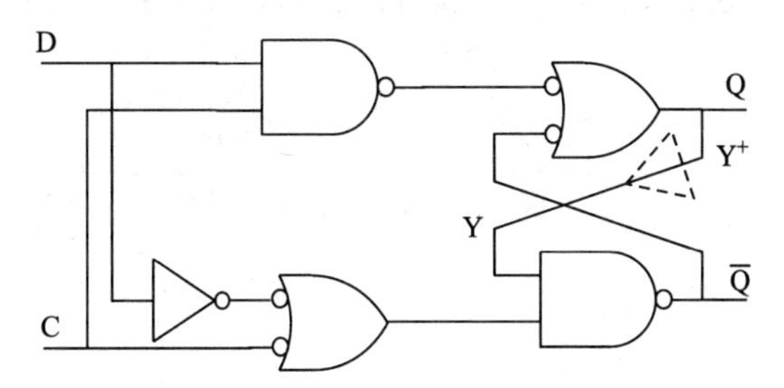

Figure 13.5 D latch with a virtual buffer.

Y	DC 00	01	11	10
0	0	0	1	0
1	1	0	1	1

Y^+

Figure 13.6 Transition table for a D latch.

We can write the state and output equations for the latch as:

$$Y^+ = D \cdot C + Y \cdot \bar{C} + D \cdot Y$$
$$Q = D \cdot C + Y \cdot \bar{C} + D \cdot Y$$
$$\bar{Q} = \bar{D} \cdot C + \bar{Y}$$

From this we can now write a *transition table* for the state variable, as shown in Figure 13.6.

A *state table* replaces the Boolean state variables with abstract states. In the state table of Figure 13.7, the stable states are circled. A state is stable when the next state is equal to the current value. The state table can also include the outputs (*state and output table*), as shown in Figure 13.7. Notice that there is an unstable state that has both outputs the same.

Using the state and output table, we can trace the change of states when an input changes. Starting from the top left corner of the table, with the current state as *K* and the two inputs at 0, let *D* change to 1. From Figure 13.8, it can be seen that the state and output remain unchanged. If *C* then changes to 1, the system moves into an unstable state. The system now has to move to the stable state at *L*, with *D* and *C* both equal to 1. Note that the state transition *must* be a vertical move on the state transition diagram. This is in order to comply with the fundamental mode restriction—anything other than a vertical move implies a change in an input value, which would therefore be occurring before the system was stable. It can be seen that the latch behaves as we would expect a D latch to behave. If *D* is changed from

S	DC 00	01	11	10
K	(K),01	(K),01	L,11	(K),01
L	(L),10	K,01	(L),10	(L),10

$S^+, Q\bar{Q}$

Figure 13.7 State table for a D latch.

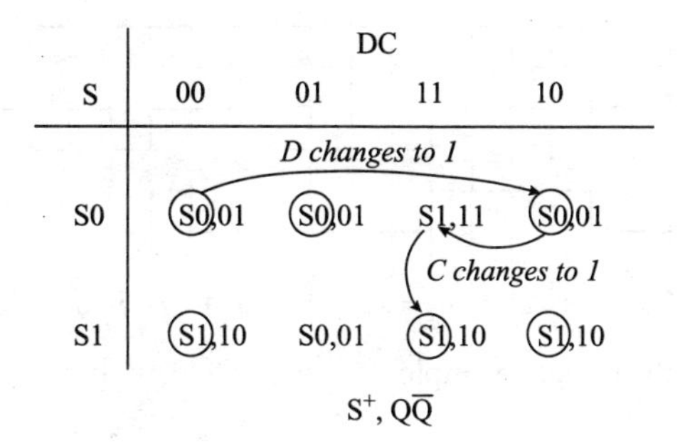

Figure 13.8 Transitions in a state table.

0 to 1, followed by *C* changing from 0 to 1, we would expect *Q* to change from 0 to 1, and it can be seen from Figure 13.8 that this is what happens.

13.3 Design of Asynchronous Circuits

In essence, the design procedure for asynchronous sequential circuits is the reverse of the analysis process. An abstract state table has to be derived, then a state assignment is performed, and finally state and output equations are generated. As will be seen, however, there are a number of pitfalls along the way, making asynchronous design much harder than synchronous design. To illustrate the procedure, we will perform the design of a simple circuit, and show, both theoretically and by simulation, the kinds of errors that can be made.

Let us design an asynchronous circuit to meet the following specification: The circuit has two inputs, *Ip* and *Enable*, and an output, *Q*. If *Enable* is high, a rising edge on *Ip* causes *Q* to go high. *Q* stays high until *Enable* goes low. While *Enable* is low, *Q* is low.

It can be see from this specification that there are eight possible combinations of inputs and outputs, but that two combinations cannot occur: if *Enable* is low, *Q* cannot be high. This leaves six states to the system, as shown in Table 13.2.

Table 13.2 States of an Example Asynchronous System

State	*Ip*	*Enable*	*Q*
a	0	0	0
b	0	1	0
c	1	0	0
d	1	1	0
e	0	1	1
f	1	1	1

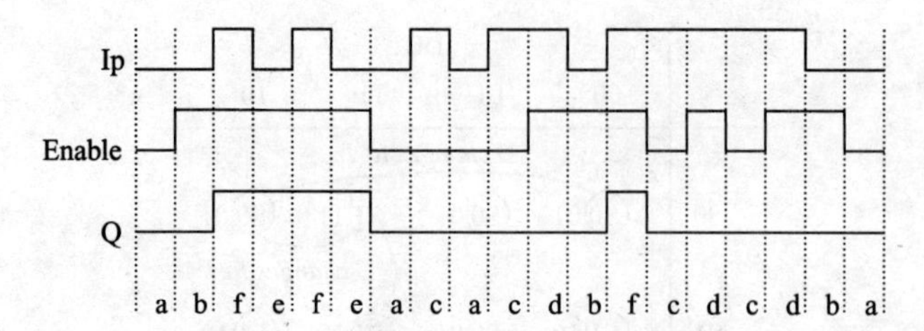

Figure 13.9 States in the design example.

The first task is to work out all the possible state transitions. One way to do this is to sketch waveforms and to mark the states as shown in Figure 13.9. From this a state transition diagram can be constructed (Figure 13.10). This state diagram can also be expressed as the *primitive flow table* of Figure 13.11. A primitive flow

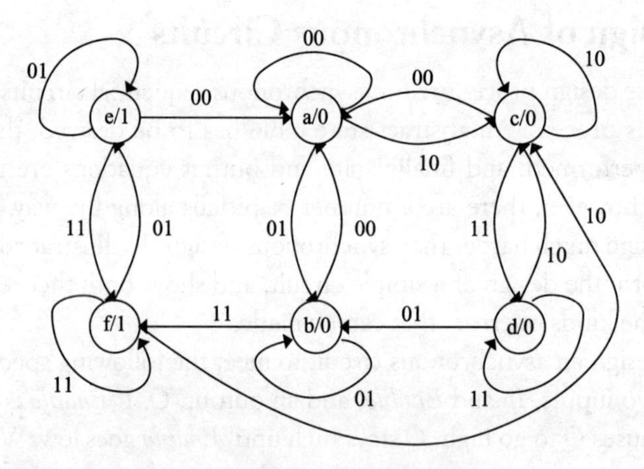

Figure 13.10 State transition diagram for the design example.

S	Ip Enable 00	01	11	10	Q
a	(a)	b	–	c	0
b	a	(b)	f	–	0
c	a	–	d	(c)	0
d	–	b	(d)	c	0
e	a	(e)	f	–	1
f	–	e	(f)	c	1
	S^+				

Figure 13.11 Primitive flow table.

table has one state per row. Because of the fundamental mode restriction, only state transitions that are reachable from a stable state with one input change are marked. State transitions that would require two or more simultaneous input changes are marked as "don't cares." The outputs are shown for the stable state and all transitions out of the state. It is also possible to assume that the outputs only apply to the stable states and that the outputs during all transitions are "don't cares."

There are more states in this primitive flow table than are needed. In Chapter 6, it was shown that states can be merged if they are equivalent. In this example, there are "don't care" conditions. We now speak of states being *compatible* if their next states and outputs are the same or "don't care." There is an important difference between equivalence and compatibility. It can be seen that states a and b are compatible and states a and c are compatible. States b and c are, however, not compatible. If a and b were *equivalent* and a and c were also equivalent, b and c would be equivalent by definition.

Here, states a and b are compatible and may be merged into state A, say. When compatible states are merged, "don't cares" are replaced by defined states or outputs (if they exist). Similarly, states c and d may be merged into C and e, and f may be merged into E. The resulting state and output table is shown in Figure 13.12.

At this point, considerable care is needed in making an appropriate state assignment. We will first demonstrate how *not* to perform a state assignment. We can show, using a SystemVerilog simulation, that a poor state assignment can easily result in a malfunctioning circuit. To encode three states requires two state variables, as described in Chapter 6. There are 24 possible state assignments. As with a synchronous system, there is no way to tell, in advance, which state assignment is "best." Therefore, let us arbitrarily assign 00 to A, 01 to C, and 11 to E. This gives the transition table shown in Figure 13.13. The state 10 is not used, so in deriving next state expressions, the entries corresponding to 10 are "don't cares."

	Ip Enable				
S	00	01	11	10	Q
A	(A)	(A)	E	C	0
C	A	A	(C)	(C)	0
E	A	(E)	(E)	C	1
		S^+			

Figure 13.12 State and output table.

	Ip Enable				
Y_1Y_0	00	01	11	10	Q
00	(00)	(00)	11	01	0
01	00	00	(01)	(01)	0
11	00	(11)	(11)	01	1
	$Y_1^+Y_0^+$				

Figure 13.13 Transition table.

Hazard-free next state and output equations can be found using K-maps:

$$Y_1^+ = Y_1 \cdot Enable + Ip \cdot Enable \cdot \bar{Y}_0$$
$$Y_0^+ = Ip + Y_1 \cdot Enable$$
$$Q = Y_1$$

A SystemVerilog model of this circuit is as follows. The next state expressions have been given arbitrary delays. It is left as an exercise for the reader to write a suitable testbench.

```
module Async_ex (input logic ip, enable,
                 output logic y0, y1);

  assign #3ns y1 = (y1 & enable) |
                   (ip & enable & ~y0);
  assign #2ns y0 = ip | (y1 & enable);
endmodule
```

If Y_1 and Y_0 are both 0 and *Ip* and *Enable* are 0 and 1, respectively, *Q* is 0. Now, let *Ip* change to 1. We would expect to move horizontally into an unstable state and then to move vertically to the stable state $Y_1Y_0 = 11$. In fact, the SystemVerilog simulation shows that the circuit goes to $Y_1Y_0 = 01$ (Figure 13.14a). If the delays are reversed, however, the circuit works as expected (Figure 13.14b):

```
assign #2ns y1 = (y1 & enable) | (ip & enable & ~y0);
assign #3ns y0 = ip | (y1 & enable);
```

Why is the circuit sensitive to these delays? We have accounted for hazards in the Boolean minimization, so they are not the problem. Let us consider the transition table, including the unused state, with the values for the unused state as implied by the minimized equations, as shown in Figure 13.15.

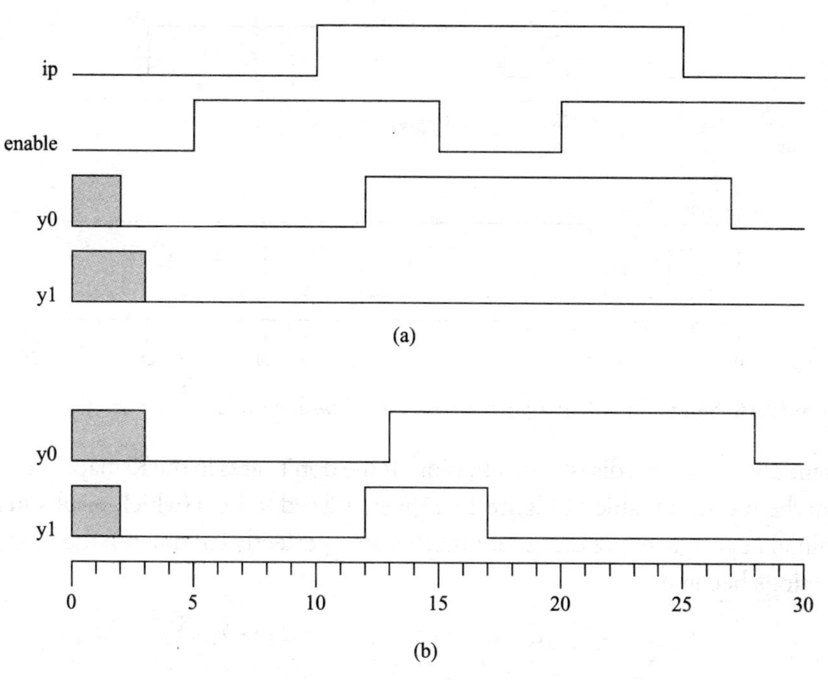

Figure 13.14 Simulation of an asynchronous circuit example: (a) with race; (b) without race.

In the first case, Y_1 changes first; therefore, the circuit changes to the unstable state 10, at which point Y_0 changes and the circuit finishes in the correct state. In the second case, Y_0 changes first and the circuit moves to the stable state 01, *and stays there*! In other words, the order in which the state variables change can affect the final state of the circuit. The situation in which two or more state variables change as a result of one input change is known as a *race*. If the final state depends on the exact order of the state variable changes, that is known as a *critical race*. There is a

	Ip Enable				
Y_1Y_0	00	01	11	10	Q
00	00	00	11	01	0
01	00	00	01	01	0
11	00	11	11	01	1
10	00	11	11	01	1
	$Y_1^+Y_0^+$				

Figure 13.15 Transition table with a critical race.

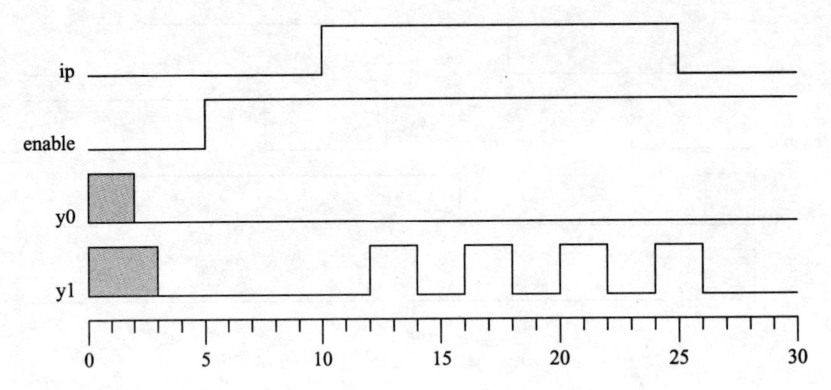

Figure 13.16 Simulation of an asynchronous circuit with a cycle.

potentially even more disastrous situation. If the don't cares in the K-maps produced from the transition table of Figure 13.13 were forced to be 0 (which results in non-minimal next state expressions, but is otherwise perfectly legitimate), the next state equations become:

$$Y_1^+ = Y_1 \cdot Y_0 \cdot Enable + Ip \cdot Enable \cdot \bar{Y}_1 \cdot \bar{Y}_0$$
$$Y_0^+ = Ip \cdot \bar{Y}_1 + I \cdot Y_0 + Y_1 \cdot Y_0 \cdot Enable \; .$$

When the SystemVerilog model shown below is simulated, the circuit oscillates, as shown in Figure 13.16.

```
assign #2ns y1 = (y1 & y0 & enable) |
                 (ip & enable & ~y1 & ~y0);
assign #3ns y0 = (ip & ~y1) | (ip & y0) |
                 (y1 & y0 & enable);
```

Figure 13.17 shows the transition table. Y_1 changes to 1 before Y_0 can react, so the circuit moves to state 10. Y_1 is then forced back to 0, so the circuit oscillates between states 00 and 01. This is known as a *cycle*.

	Ip Enable				
Y_1Y_0	00	01	11	10	Q
00	00	00	11	01	0
01	00	00	01	01	0
11	00	11	11	01	1
10	00	00	00	00	1
	$Y_1^+Y_0^+$				

Figure 13.17 Transition table with a cycle.

S	00	01	11	10	Q
	Ip Enable				
A	(A)	(A)	G	C	0
C	A	A	(C)	(C)	0
E	G	(E)	(E)	C	1
G	A	–	E	–	–
			S^+		

Figure 13.18 Modified state table.

We clearly have to perform a state assignment that avoids both critical races and cycles. In this example, such an assignment is not possible with just three states. Therefore, we have to introduce a fourth state. This state is unstable, but it ensures that only one state variable can change at a time. Figure 13.18 shows the modified state table, while Figure 13.19 shows a simplified state transition diagram, with the newly introduced state, G, and a suitable state assignment. Hence, expressions for the state variables can be derived. In this case, the state variable expressions are:

$$Y_1^+ = Y_1 \cdot Y_0 \cdot \bar{I} + Y_1 \cdot Enable + Ip \cdot Enable \cdot \bar{Y}_0$$
$$Y_0^+ = Ip \cdot \overline{Enable} + Ip \cdot Y_0 + Y_1 \cdot Enable$$

We can simulate SystemVerilog models of this circuit with either Y_1 or Y_0 changing first, and in both cases the circuit works correctly.

There is, however, one final potential problem. There are no possible redundant terms in this example, so we can be sure that all potential static hazards have been eliminated. In principle, therefore, the circuit can be built as shown in Figure 13.20. If, however, as a result of the particular technology used or the particular layout adopted, the input to the top AND gate is delayed with respect to the state variables, as shown, the circuit may still malfunction. This condition can be demonstrated again with a SystemVerilog model.

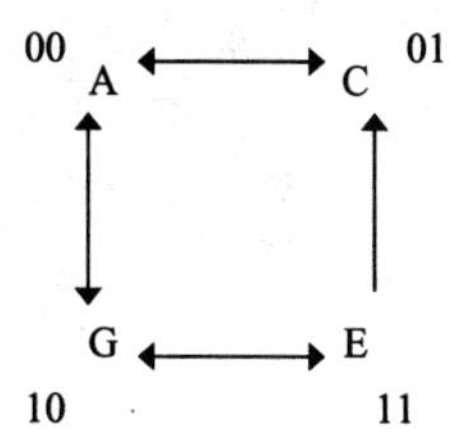

Figure 13.19 Simplified state transition diagram.

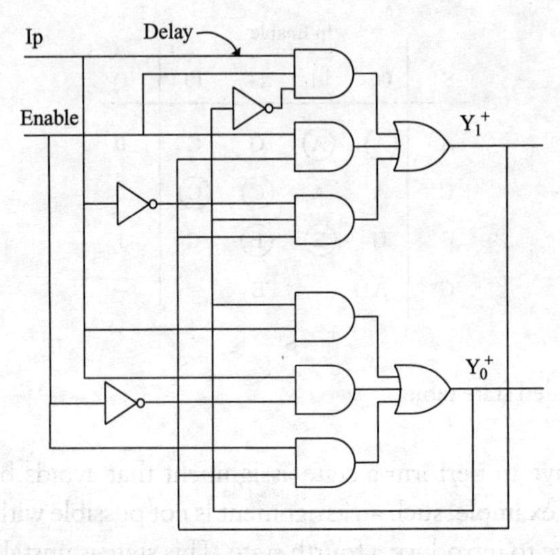

Figure 13.20 Circuit with an essential hazard.

```
assign #5ns ipslow = ip;
assign #2ns y1 = (y1 & y0 & ~ip) | (y1 & enable) |
                 (ipslow & enable & ~y0);
assign #3ns y0 = (ip & y0) | (ip & ~enable) |
                 (y1 & enable);
```

The transition table of Figure 13.21 shows what happens if *Ip* changes from 1 to 0 from state 01 while *Enable* stays at 1. In theory, this change should only cause transitions 1*a* and 1*b* and the final state should be 00. In practice, because of the delay in *Ip*, the circuit then follows the other transitions shown, 2*a*, 2*b*, 3*a*, and 3*b*, to finish in state 11. This is known as an *essential hazard*, so-called because it is part of the essence of the circuit. Potential essential hazards can be identified from the transition table if a single input change results in a different final state than if the

Y_1Y_0 \ Ip Enable	00	01	11	10	Q
00	(00)	(00)	10	01	0
01	00	00	(01)	(01)	0
11	10	(11)	(11)	01	1
10	00	–	11	–	–

Figure 13.21 Transition table with an essential hazard.

input changes three times. The only way to avoid essential hazards is to ensure that the state variables cannot be fed back around the circuit before the input transitions. This can be achieved by careful layout or possibly by deliberately introducing delays into the state variables.

In summary, therefore, the design of an asynchronous sequential circuit has the following steps.

1. State the design specifications.
2. Derive a primitive flow table.
3. Minimize the flow table.
4. Make a race-free state assignment.
5. Obtain the transition table and output map.
6. Obtain hazard-free state equations.
7. Check for essential hazards.

13.4 Asynchronous State Machines

In the design flow, the first step is to derive the design specifications. In many ways this is the hardest part of the task. Moreover, if we get that wrong, everything that follows is also, by definition, wrong. By the nature of the design process, it is almost impossible to patch a mistake—the entire process has to be repeated. Therefore, it would be very desirable to ensure that the design has been specified correctly. One way to do this is to use simulation again.

The state transition diagram of Figure 13.10 is essentially the same as the state diagram of Figure 6.8 or that of Figure 12.19. Figure 13.10 represents an asynchronous system and the other two represent synchronous systems. This difference is not, however, apparent from the diagrams. We advocated the use of ASM charts for the design of synchronous systems, but we could have used state diagrams. We know that an ASM chart or a state diagram has an equivalent SystemVerilog description. By the same argument, we can represent an asynchronous state machine in SystemVerilog. Instead of a set of registers synchronized to a clock, we would have a virtual buffer, in which the state variable is updated. Therefore, let us write a SystemVerilog description of the state machine of Figure 13.10.

```
module async_sm (input logic ip, enable,
                 output logic q);

enum {a, b, c, d, e, f} present_state, next_state;

always @*
```

```
  begin
  next_state = present_state;
  q = '0;

  case (present_state)
    a: if (!ip && enable)
         next_state = b;
       else if (ip && !enable)
         next_state = c;

    b: if (!ip && !enable)
         next_state = a;
       else if (ip && !enable)
         next_state = f;

    c: if (!ip && !enable)
         next_state = a;
       else if (ip && enable)
         next_state = d;

    d: if (!ip && enable)
         next_state = b;
       else if (ip && !enable)
         next_state = c;

    e: begin
       q = '1;
       if (!ip && !enable)
         next_state = a;
       else if (ip && enable)
         next_state = f;
       end

    f: begin
       q = '1;
       if (!ip && enable)
         next_state = e;
       else if (ip && !enable)
         next_state = c;
       end
  endcase
  end

  assign #1ns present_state = next_state;

endmodule
```

The virtual buffer has a delay of 1 ns. For this type of model to work, there must be a finite delay—a zero delay would cause the process to loop infinitely at time 0. For reasons of space, the entire state machine is not shown; the other states may be written in the same way. The don't cares have been assumed to cause the state machine to stay in the same state. As these represent violations of the fundamental mode, this is valid. With a suitable testbench, we can use this SystemVerilog model to reproduce Figure 13.9. Notice that the initial values of the state variables will be the leftmost entry in the state definition—a.

We can also repeat the exercise after state minimization.

```
module async_smr (input logic ip, enable,
                  output logic q);

enum {A, C, E} present_state, next_state;

always @*
  begin
  next_state = present_state;
  q = '0;

  case (present_state)
    A: if (ip && enable)
         next_state = E;
       else if (ip && !enable)
         next_state = C;

    C: if (!ip)
         next_state = A;

    E: begin
       q = '1;
       if (!ip && !enable)
         next_state = A;
       else if (ip && !enable)
         next_state = C;
       end
  endcase
  end

  assign #1ns present_state = next_state;

endmodule
```

Again, this can be verified by simulation. Indeed, this is one way to check that the state minimization has been done correctly.

As a second example, consider the following. We wish to design a phase detector with two outputs: *qA* and *qB*. There are also two inputs: *inA* and *inB*. Let us assume both outputs start high. When *inA* goes high, *qA* goes low and stays low until *inB* goes high. Similarly, if *inB* goes low first, *qB* goes low until *inA* goes high. This sounds very simple! We will model the phase detector as an asynchronous state machine. It is left as an exercise for the reader to derive the SystemVerilog model to implement this specification. You can further test your understanding of asynchronous design by taking this design through to the gate level.

```
module phase_detector (input logic inA, inB,
                       output logic qA, qB);

enum {A, B, C, D, E, F, G, H} present_state,
                              next_state;

always @*
  begin
  next_state = present_state;
  qA = '1;
  qB = '1;

  case (present_state)
    A: if (~inA && inB)
         next_state = E;
       else if (inA && ~inB)
         next_state = B;
    B: begin
       qA = '0;
       if (~inA && inB)
         next_state = D;
       else if (inA && inB)
         next_state = C;
       end
    C: if (~inA && inB)
         next_state = D;
       else if (inA && ~inB)
         next_state = F;
    D: if (~inA && ~inB)
         next_state = A;
       else if (inA && inB)
         next_state = H;
    E: begin
       qB = '0;
       if (inA && inB)
         next_state = C;
```

```
      else if (inA && ~inB)
        next_state = F;
      end
   F: if (~inA && ~inB)
        next_state = A;
      else if (inA && inB)
        next_state = G;
   G: begin
      qB = '0;
      if (~inA && ~inB)
        next_state = E;
      else if (~inA && inB)
        next_state = D;
      end
   H: begin
      qA = '0;
      if (~inA && ~inB)
        next_state = B;
      else if (inA && ~inB)
        next_state = F;
      end
  endcase
  end

  assign #1ns present_state = next_state;

endmodule
```

One final word of warning: Do not try to synthesize these state machine models! In the light of the previous discussions, it should be obvious that you would generate hardware with races and hazards. We have used **always** @* rather than **always_comb** to show that these are not synthesizable models.

13.5 Setup and Hold Times and Metastability

13.5.1 The Fundamental Mode Restriction and Synchronous Circuits

The fundamental mode restriction requires that an input to an asynchronous circuit must not change until the circuit has become stable after a previous input change. Individual flip-flops are themselves asynchronous internally, but are used as synchronous building blocks. We do not, however, speak of the fundamental mode restriction when designing synchronous systems. Instead, we define setup and hold times.

Because of the gate delays in a circuit, the fundamental mode restriction *does not* mean that two inputs must not change at the exact same time. It means that the effect of one input change must have propagated through the circuit before the next input can change. To use the example of a D flip-flop, a change at the D input must have propagated through the flip-flop before an active clock edge may occur. Similarly, the effect of the clock edge must have propagated through the circuit before the D input can change again. These two time intervals are known as the setup and hold times, respectively.

The setup and hold times of a latch or flip-flop depend on the propagation delays of its gates. These propagation delays depend, in turn, on parametric variations. So we can never know the exact setup and hold times of a given flip-flop. Furthermore, the timing of clock edges may be subject to *jitter*—the exact period of the clock may vary slightly. Therefore, there has to be a margin of tolerance in estimating the setup and hold times. It should finally be noted that some of the effects of ignoring the fundamental mode restriction, or equivalently, violating setup and hold times, are not purely digital. In particular, metastability is effectively an analog phenomenon.

Bearing all this in mind, it is possible to get some insight into the consequences of not observing the fundamental mode restriction by using a SystemVerilog simulator. We can use **`specify`** blocks, introduced in Chapter 10, to verify timing behavior.

13.5.2 SystemVerilog Modeling of Setup and Hold Time Violations

A structural model of a level-sensitive D latch can be described in SystemVerilog using gate instances, as shown in the following. If a simulation of this latch is run using a regular clock and a random event generator for the D input, as shown in the testbench fragment, it will be observed that the latch works correctly, unless the D input changes from 0 to 1 in the interval 2 ns or less before a falling clock edge. If this occurs, the q and qbar outputs oscillate.

Of course, two D latches can be put together to form an edge-triggered flip-flop. The clock input is inverted for the master flip-flop (introducing a delay of, say, 1 ns). Thus, when the clock is low, the master flip-flop is transparent. From the previous simulation, we would expect that the setup time is 2 ns, less the delay in the clock caused by the inverter, or 1 ns in total. We can verify this by simulation. Again, we observe that a change in the D input 1 ns or less before the clock edge may cause the output to oscillate, depending on the state of the flip-flop and whether D is rising or falling. The six-nand gate edge-triggered D flip-flop behaves similarly. In both cases, the hold time is 0 ns.

```
module dlatchnet(output wire q, qbar,
                 input wire d, c);

  wire e, f, g;

  not #1ns g0 (e, d);
  nand #1ns g1 (f, d, c);
  nand #1ns g2 (g, e, c);
  nand #1ns g3 (qbar, g, q);
  nand #1ns g4 (q, f, qbar);

endmodule
```

The testbench follows. The `$dist_exponential` system function is used to specify a randomized delay. In this function, the mean time to the next event is specified, but instead of a uniform distribution, half the event times will be between zero and the mean time, and half will be between the mean time and infinity. This model is commonly used in queuing theory. Because this function takes an integer, the **timeunit** and **timeprecision** are declared at the start of the testbench. A seed is used to ensure a different value is returned each time.

```
module testdlatch;

  timeunit 1ns;
  timeprecision 100ps;

  logic q, qbar, d, c;
  int seed;

  dlatchnet d0 (.*);

  initial
    begin
    d = '0;
    forever #($dist_exponential(seed, 20)) d = ~d;
    end

  initial
    begin
    c = '0;
    forever #10 c = ~c;
    end

endmodule
```

Warnings about setup time violations can be generated by including a **$setup** system call within a **specify** block.

```
specify
  $setup(d, negedge c, 2ns);
endspecify
```

Similarly, hold times can be checked with a **$hold** system call. Note that the clock and d inputs would be reversed in the **$hold** call.

There has to be some doubt as to whether this modeled behavior is exactly what would be observed in a real circuit. These SystemVerilog models assume that 0 to 1 and 1 to 0 transitions are instantaneous. Of course, in reality, such transitions are finite. Therefore, if a gate had one of its two inputs rising and the other falling simultaneously, it would be reasonable to expect that the output might switch into some state that was neither a logic 1 nor a logic 0 for a period of time. SystemVerilog does not include such a state; "x" is generally taken to represent a state that could be one of 1 or 0, but not neither.

13.5.3 Metastability

While the oscillations predicted by both structural models may occur if the fundamental mode restriction is violated, another condition can occur that a SystemVerilog simulation cannot predict. All flip-flops have two stable states and a third unstable, or *metastable*, state. In this metastable state, both flip-flop outputs have an equal value at a voltage level between 0 and 1. A SPICE, or similar transistor-level operating point analysis is likely to find this metastable condition. This may be likened to balancing a pencil on its point—in theory, it is stable, but in practice, noise (vibrations, air movement, etc.) would cause the pencil to topple. The metastable state of a flip-flop is similarly unstable; electrical or thermal noise would cause it to fall into a stable state.

Metastability is most likely to occur when external (asynchronous) signals are inputs to a synchronous system. If metastability is likely to be a problem, then care needs to be taken to minimize its effects. The threat of metastability can never be entirely eliminated, but there is no point in constructing elaborate defenses if the chances of its happening are remote. Therefore, the critical question is how likely is it to occur? The formula used to calculate the mean time between failures (MTBF) has been found, by experiment, to be:

$$MTBF = \frac{\exp(T \times t_x)}{f_{clk} \times f_{in} \times T_0}$$

t_x is the time for which metastability must exist in order for a system failure to occur. If a metastable state occurs at the output of a flip-flop, it will cause a problem if it propagates through combinational logic and affects another flip-flip. Therefore,

$$t_x = t_{clk} - t_{pd} - t_{setup}$$

where t_{clk} is the clock period, t_{pd} is the propagation delay through any combinational logic, and t_{setup} is the setup time of the second flip-flop.

f_{clk} is the clock frequency, f_{in} is the frequency of the asynchronous input changes, and T and T_0 are experimentally derived constants for a particular device.

Let us put some numbers into this formula. The system is clocked at 10MHz; therefore, t_{clk} is 100 ns. We will examine whether an input flip-flop with a setup time of 10 ns can go into a metastable state; therefore, t_{pd} is zero and, hence, t_x is 90 ns. If the asynchronous input changes on average, say, once every 10 clock cycles, f_{in} is 1MHz. For a relatively slow D flip-flop (e.g., a 74LS74), T is about 7×10^8 sec, while T_0 is 0.4 sec. Therefore,

$$MTBF = \frac{\exp(7 \times 10^8 \times 90 \times 10^{-9})}{10^7 \times 10^6 \times 0.4} = 5.7 \times 10^{12}\,\text{sec}$$

or about 200,000 years. Metastability is unlikely to be a problem in such a system. But suppose the clock frequency is doubled to 20MHz, and hence t_x becomes 40 ns. Now,

$$MTBF = \frac{\exp(7 \times 10^8 \times 40 \times 10^{-9})}{2 \times 10^7 \times 10^6 \times 0.4} = 0.18\,\text{sec.}$$

So, we probably will have a problem with metastability in this system.

There are several ways to alleviate the problem. The flip-flop cited above is very slow. A faster flip-flop would have a larger T and a smaller T_0. So, using a faster flip-flop will increase the MTBF. Another common solution is to use two flip-flops in series as shown in Figure 13.22.

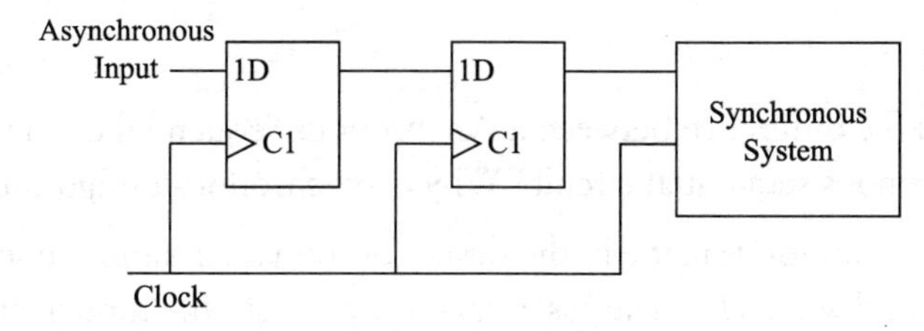

Figure 13.22 Synchronizer design.

This arrangement does not necessarily reduce the MTBF, but it does reduce the possibility that a metastable state is propagated into the synchronous system.

Although it is fairly unlikely that metastability would be observed in a student laboratory, it is apparent that with increasing clock speeds and perhaps a move toward a style of design in which there is no global clock, coping with metastability is going to be a challenge for digital designers.

Summary

The design and analysis of asynchronous circuits is harder than for synchronous circuits. Asynchronous circuits may be formally analyzed by breaking feedback loops. The design of an asynchronous circuit starts from a description of all the possible states of the system. A primitive flow table is constructed, which is then minimized. State assignment follows. A poor state assignment can result in race conditions or cycles. From the transition table, next state and output expressions are derived. Hazards can cause erroneous behavior or oscillations. Essential hazards may result from uneven delays. The design of asynchronous circuits depends on observing the fundamental mode restriction. This is reflected in the specification of setup and hold times for asynchronous blocks used in synchronous design. Failure to observe these restrictions can lead to spurious behavior and possibly metastability.

Further Reading

Although the design of asynchronous (or level-mode or fundamental mode) sequential circuits is covered in many textbooks, close reading reveals subtle variations in the techniques. Hill and Peterson [10] provide a very good description. Wakerly [25] has a very straightforward explanation. Unger's 1995 paper [23] has provided perhaps the most rigorous analysis of the problems of metastability. The Amulet project has one of the most significant large asynchronous designs and the Web site (`www.cs.man.ac.uk/amulet/index.html`) has links to many sources of information about asynchronous design.

Exercises

13.1 What is the difference between a synchronous sequential circuit and an asynchronous sequential circuit? Why is synchronous design preferred?

13.2 What assumption is made in the design of fundamental-mode sequential circuits and why? How can essential hazards cause the fundamental mode to be violated?

13.3 The excitation equation for a D latch may be written as

$$Q^{+} = C \cdot D + Q \cdot \bar{C}$$

Why would a D latch implemented directly from this transition equation be unreliable? How would the D latch be modified to make it reliable?

13.4 Describe, briefly, the steps needed to design an asynchronous sequential circuit.

13.5 Figure 13.23 shows a master-slave edge-triggered D flip-flop. How many feedback loops are there in the circuit, and hence how many state variables? Derive excitation and output equations and construct a transition table. Identify all races and decide if the races are critical or non-critical. Construct a state and output table and show that the circuit behaves as a positive edge-triggered flip-flop.

13.6 Figure 13.24 shows a state diagram of an asynchronous circuit with two inputs, R and P, and a single output, Q. The input values are shown on the arcs; the state names and the output values of the stable states are shown in the circles. Design an asynchronous circuit to implement this function.

13.7 A positive edge-triggered D flip-flop has set and reset inputs, in addition to the clock and D inputs (Figure 13.4). Write down the state equations for the flip-flop including the set and reset inputs. Hence, write a transition table.

13.8 Table 13.3 shows the transition table for an asynchronous circuit. Identify all the non-critical races, critical races, and cycles (a *cycle* is a repeated series of unstable states that requires an input to change in order for a stable state to be reached).

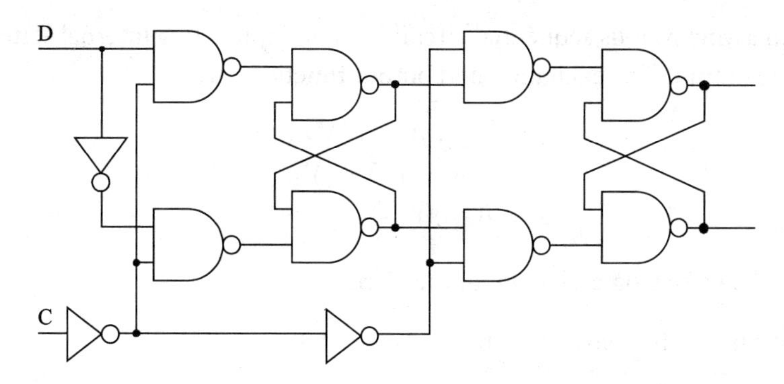

Figure 13.23 Circuit for Exercise 13.5.

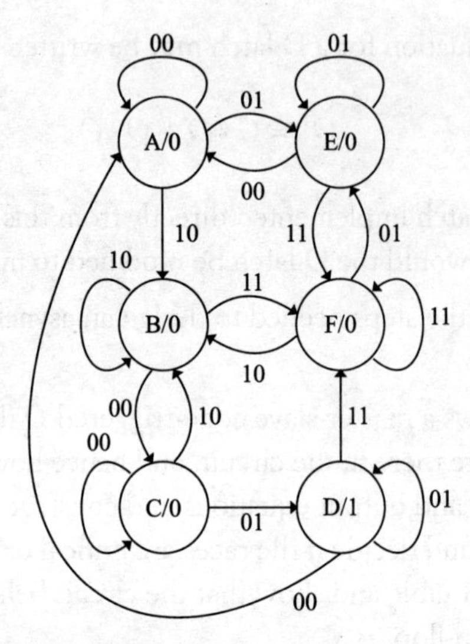

Figure 13.24 State diagram for Exercise 13.6.

Table 13.3 Transition Table for Exercise 13.8

	AB			
*Y*1*Y*2	00	01	11	10
00	00	11	10	11
01	11	01	01	10
11	10	11	01	10
10	11	10	01	01
		Y1*Y2*		

13.9 Design a D flip-flop that triggers on both the positive and negative edges of the clock pulse.

13.10 An asynchronous sequential circuit has two inputs, two internal states, and one output. The excitation and output functions are:

$$Y1^{+} = A \cdot B + A \cdot \overline{Y2} + \bar{B} \cdot Y1$$
$$Y2^{+} = B + A \cdot \overline{Y1} \cdot \overline{Y2} + \bar{A} \cdot Y1$$
$$Z = B + Y1$$

(a) Draw the logic diagram of the circuit.

(b) Derive the transition table and output map.

(c) Obtain a flow table for the circuit.

14 Interfacing with the Analog World

In previous chapters, we considered the world to be purely digital. Indeed, with the exception of Chapter 13, we further considered only synchronous systems. Of course the real world is asynchronous and, even worse, analog. All digital systems must at some point interact with the real world. In this chapter, we consider how analog inputs are converted to digital signals and how digital signals are converted to analog outputs. Until relatively recently, the modeling and simulation of digital and analog circuits and systems would have been performed independently of each other. A set of analog and mixed-signal extensions to Verilog (but not yet SystemVerilog) has been proposed. The language is commonly known as Verilog-AMS (analog and mixed-signal). Verilog-AMS is a complete superset of the 2005 standard for Verilog. At some point in the future, it is likely that SystemVerilog-AMS will appear, but meanwhile, simulators that support mixtures of Verilog, Verilog-AMS, and SystemVerilog exist.

Having looked at digital-to-analog converters (DACs) and analog-to-digital converters (ADCs), we will review the basics of Verilog-AMS and see how ADCs and DACs can be modeled in Verilog-AMS. There is insufficient space to provide a complete tutorial of Verilog-AMS here. Furthermore, it should be remembered that we are only considering simulation models, designed for verifying the interaction of a digital model with the real world. Synthesis of analog and mixed-signal designs is still a research topic. The final section of this chapter looks at some further mixed-signal circuits and their models in Verilog-AMS.

14.1 Digital-to-Analog Converters

We start the discussion of interface circuits with DACs because, as we will see, one form of ADC requires the use of a DAC. The motivation in this chapter is not to describe every possible type of converter—that would require at least an entire book—but to show one or two examples of the type of circuit that can be employed.

In moving between the analog and digital worlds, we ideally want to preserve the maximum amount of information. This can be summarized in terms of three aspects: *resolution*, *accuracy*, and *speed*. *Resolution* defines the smallest change that can be measured. For example, 8 bits can represent 2^8 or 256 voltage levels. If we want to represent a signal that changes between 0 V and 5 V using 8 bits, the resolution is 5/256 = 19.5 mV. *Accuracy* describes how precisely a signal is represented with respect to some reference. In turn, this depends on factors such as linearity. For example, while 8 bits can represent a 5 V signal with an *average* resolution of 19.5 mV, differences (non-linearities) in the circuit might mean that some changes are really 18.5 mV, while others are 20.5 mV. These differences will add up and affect the overall accuracy. Finally, the *speed* at which data is converted between the two domains affects the design of converters. In the digital world, samples are taken at discrete points in time. The users of converters need to be aware of what happens between these sample points.

The simplest type of DAC is the binary-weighted ladder circuit of Figure 14.1. The bits are added together according to their relative weights. The operational amplifier forms a classic (inverting) adder. While this circuit is easy to understand, it is difficult to manufacture. The resistors have to be made with very tight tolerances. Any inaccuracy in a resistor value would affect the accuracy. Note that the resistors have to be accurate with respect to the feedback resistor (R), but also with respect to each other.

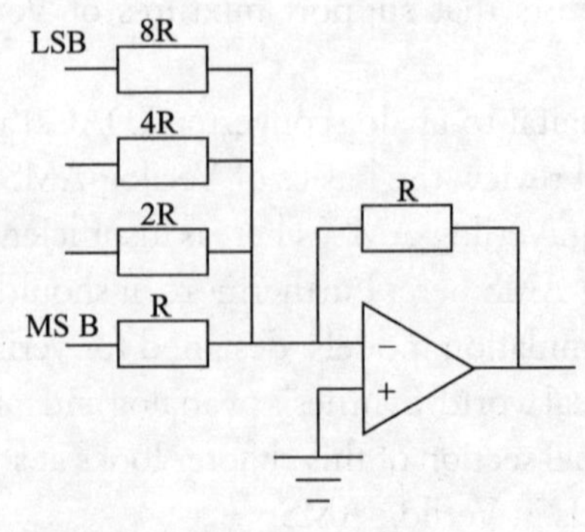

Figure 14.1 Binary-weighted ladder DAC.

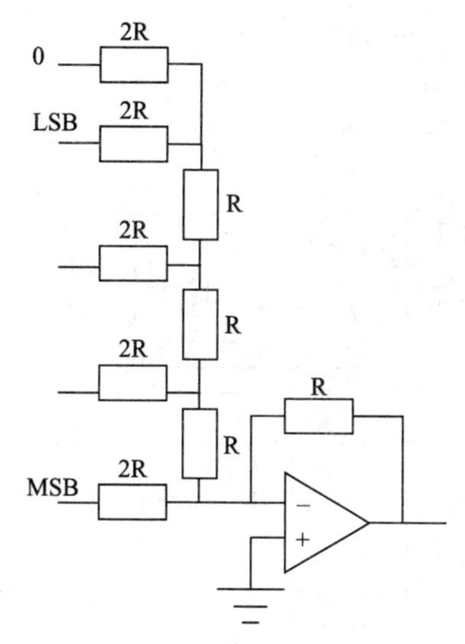

Figure 14.2 Binary-weighted R-2R ladder DAC.

A variation on the binary-weighted ladder is the R-2R ladder of Figure 14.2. To a significant extent, this overcomes the manufacturing problem as only two values of resistor need to be constructed.

For both these circuits, the speed is limited only by the response of the opamp. In practice, however, we might find that the resistors are more easily implemented as switched capacitors.[1] If this is so, the speed is limited by the clock. Notice also that the output changes in discrete steps.

14.2 Analog-to-Digital Converters

The task of an ADC is to translate a voltage (or current) into a digital code. This is generally harder to achieve than the reverse process. Again, we need to consider resolution, accuracy, and speed, but for example, suppose we have a signal that changes between 0 V and 5 V, with a maximum frequency of 10kHz. Eight bits gives a resolution of 19.5 mV. To accurately capture changes in a signal, it needs to be sampled at twice its maximum frequency. Here, therefore, we need to sample at 20kHz or greater.

1. In CMOS technology, it is generally easier to build matched capacitors than matched resistors. It is possible to emulate the behavior of a resistor by rapidly switching a capacitor between an input and ground.

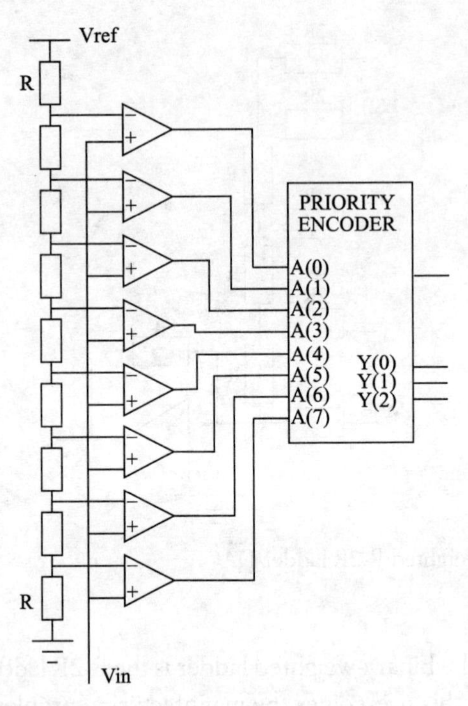

Figure 14.3 Flash ADC.

Conceptually, the simplest ADC is the flash ADC of Figure 14.3. This consists of nine identical resistors (for eight voltage levels) and eight comparators. As the input voltage, Vin, increases past a level in the resistor chain, the corresponding comparator output switches to 1. Therefore, we can use a priority encoder to determine which is the most significant bit, and to encode that value as a binary number. It should be immediately obvious that this circuit is impractical for large numbers of bits. We need 2^n identical, ideal comparators and $2^n + 1$ identical resistors to achieve n bits at the output. It is very difficult to achieve high consistency and hence high accuracy. On the other hand, this type of converter is very fast. In practice, the cost of a flash ADC is usually too high. In return for a smaller design and better accuracy, we pay the price of slower conversion speeds.

Figure 14.4 shows a tracking ADC. This is much easier to implement than the flash ADC. It is essentially a DAC, a comparator, and a counter. When the value in the counter is greater than that of the input, Ain, the counter counts down; when the counter's value is less than Ain, the counter counts up. Therefore, the counter attempts to track the input. As might be expected, a very high clock speed is needed to make this work. Suppose we wish to convert an audio signal with a maximum frequency of 20kHz. We need to sample at twice this frequency—40kHz.

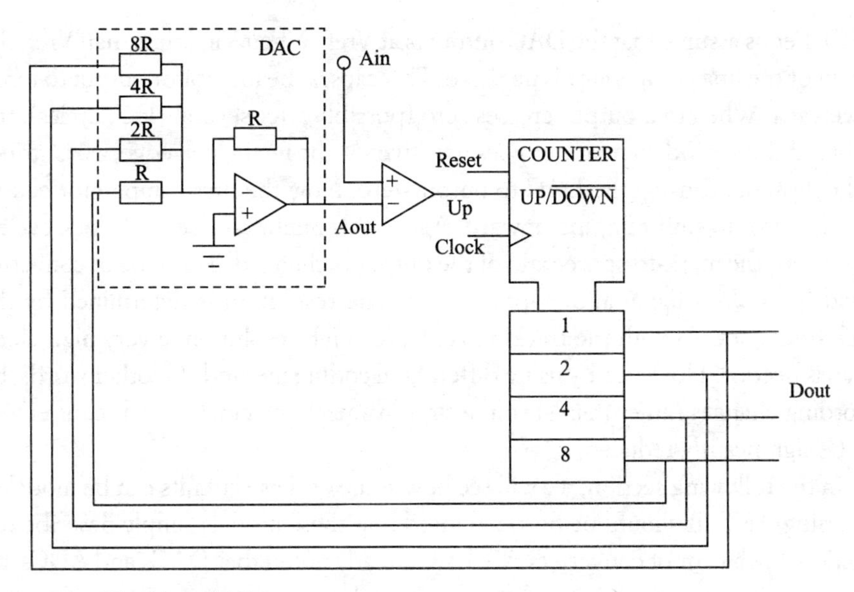

Figure 14.4 Tracking ADC.

In the worst case, the counter needs to count through its entire range, 2^4 or 16 states, between samples. This means that the counter clock must be 16 × 40kHz or 640kHz. On the other hand, to achieve CD quality resolution, we would need 16 bits at the output, which implies a clock speed of nearly 3GHz. This is clearly much less practical.

For high-speed, high-resolution applications, an entirely different approach is usually taken. Delta-sigma ADCs convert from voltage to a serial encoding. Figure 14.5 shows a simple delta-sigma ADC. The mark-to-space ratio of the output is proportional to the ratio of the input voltage to some reference, Vref, (as set by the

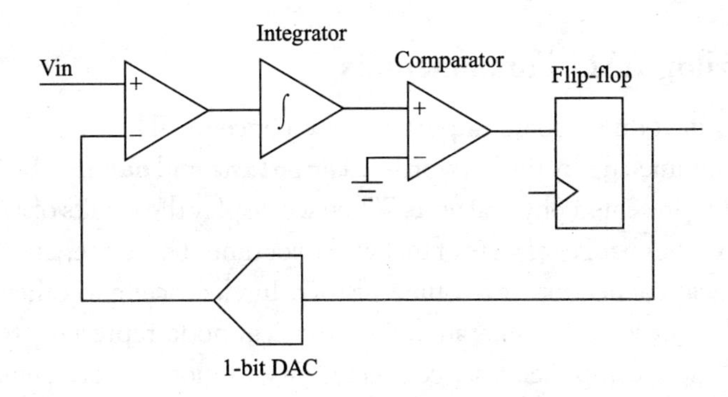

Figure 14.5 Delta-sigma ADC.

DAC). Let us assume that the DAC output is at Vref. When Vin is less than Vref, the output of the first comparator is negative. This causes the integrator output to ramp downward. When that output crosses zero (possibly after several clock cycles), the output of the second comparator goes negative. At the next clock edge, a 0 is stored in the flip-flop, causing the DAC to output zero. Now the first comparator causes the integrator to start ramping upward. Again, this might take several clock cycles. In this way, the mark-to-space ratio of the output is changed. This type of converter is widely used in digital audio applications. The resolution is determined by the clock frequency. As with the tracking ADC, for high resolution, a very high clock speed is needed. However, by using differential coding methods (in other words, by recording changes rather than absolute signal values), the clock speed requirement can be significantly reduced.

In the following section, we will see how some of these circuits can be modeled in Verilog-AMS. It should be borne in mind that these models simply describe the functional behavior of converters. We have already noted that DACs and ADCs are subject to limitations in terms of accuracy, resolution, and speed. Very often it is necessary to model these imperfections and to use the results of such simulations to determine the most suitable designs. As with much else in this chapter, detailed modeling of converter circuits could comprise yet another complete book.

14.3 Verilog-AMS

Verilog-AMS is a superset of Verilog (2005) (see Appendix). Several new keywords and constructs have been added to allow modeling of physical systems. The standard defines the interaction between a standard Verilog simulator and an analog solver. It is important to realize that Verilog-AMS is not "analog Verilog", but a true mixed-signal modeling language. Moreover, Verilog-AMS has been designed to allow modeling of general physical systems, not simply electrical networks.

14.3.1 Verilog-AMS Fundamentals

Verilog-AMS introduces some important new concepts. The most important of these can be summed up by the keywords: **discipline** and **nature**. In "standard" Verilog, a net represents a physical wire. When we display the results of a simulation, we can observe the changes of state of that wire over time. Therefore, a net covers two ideas: a physical connection and a time history. In electrical and other networks, these two ideas need to be separated. An electrical node represents the point at which two or more components are connected. We cannot, however, talk about the behavior of that node, unless we specify whether we are referring to its voltage or current or some other aspect.

To distinguish between the physical connection and the behavior, Verilog-AMS introduces new keywords. The voltage or current at a node is declared as a **nature**. Before giving an example, however, we need to explain how natures and nodes relate to each other.

A node belongs to a particular type of network. For example, an electrical node belongs to an electrical network; a magnetic node belongs to a magnetic network. Each type of network has behavior that can be described in terms of natures. So, for example, the behavior of an electrical network can be described in terms of voltages and currents, while the behavior of a magnetic network can be described in terms of magneto-motive force and flux. Each type of network has a pair of natures. These can be described as **flow** and **potential** natures. For example, in an electrical network, current flows from one node to another through network components, while we can also measure the voltage potential across such components. Each type of network has such a pair of flow and potential natures. (Note that it is also possible to define an electrical network in which currents are thought of as the potential natures and voltages are thought of as the flow natures. Mathematically, either convention is acceptable. The first convention is more common, however, and we will stick with that. In other kinds of networks, the decision about which quantity is potential and which is flow may be less clear.)

In declaring that a node belongs to a particular kind of network, we are effectively defining the flow and potential natures for that net. Therefore, it would not be adequate to declare a net to be of a particular Verilog type. Instead, a new construct is used—a **discipline**. A discipline has two parts: a potential nature and a flow nature. Additionally, a **domain** can be declared: **continuous** or **discrete**. By default, disciplines are continuous.

An electrical discipline might be declared as:

```
discipline electrical
  potential Voltage;
  flow Current;
enddiscipline
```

As elsewhere, the keywords are highlighted. What are Voltage and Current? We know that the potential and flow parts are natures, so Voltage and Current must be natures:

```
nature Voltage
  units = "V";
  access = V;
  idt_nature = Flux;
  abstol = 1e-6;
endnature
```

```
nature Current
  units = "A";
  access = I;
  idt_nature = Charge;
  abstol = 1e-12;
endnature
```

Each of these declarations has four parts, but not all need to be declared. First, **units** shows the symbol used for that nature. Verilog-AMS does not perform dimensional analysis, so this is simply to provide readability. Second, an **access** function is given. This allows, for example, the voltage at node1 to be referenced by writing `V(node1)`. The third part, **idt_nature**, shows the nature that results when this quantity is integrated with respect to time. It is also possible to declare the time derivative, **ddt_nature**. In this case, Flux and Charge would be declared elsewhere. Finally, the **abstol** or absolute tolerance is declared. This defines the accuracy to which variables of that nature should be calculated.

In the examples that follow, we will assume that the definitions of the electrical discipline are contained in a file, `disciplines.vams`, that is included at the start of each model. We can now define one or more nodes with the electrical discipline:

```
electrical node1, node2;
```

Nodes can be declared within modules (in exactly the same way as nets) or as ports. In port declarations, the direction must be specified as **inout**. For example, the module declaration of a resistor might be:[2]

```
`include "disciplines.vams"
module resistor (node1, node2);
  inout node1, node2;
  electrical node1, node2;
  parameter real R = 1;
```

At this point, we have only created the physical nodes. We can refer to the current flowing between those nodes by `I(node1, node2)` or we can declare one or more branches. A branch between the two nodes would be declared as:

```
branch (node1, node2) res;
```

2. Note that this is written in the Verilog 1995 style. The latest version of the standard supports the Verilog 2005/SystemVerilog style of headers, but at the time of writing, this is not supported by any simulators.

Now the current can be referred to as `I(res)`.

The reference node is the name of the terminal with respect to which all across quantities are calculated. In electrical networks, this is often known as the ground or earth node. In a Verilog-AMS model, this can be declared with:

```
ground gnd;
```

14.3.2 Contribution Statements

Contribution statements define the network equations of analog models. Contribution statements use the symbol "< +" to show how an expression contributes to the overall set of network equations. Note that this is not an assignment in the conventional sense, as multiple contributions to the same flow or potential are summed.

The contribution statements are therefore the simultaneous equations that are solved by the analog simulator. Contribution statements must be put within an **analog** procedural block. In order to illustrate a contribution statement, we will give a complete model of a resistor:

```
`include "disciplines.vams"
module resistor (node1, node2);
  inout node1, node2;
  electrical node1, node2;
  parameter real R = 1;
  branch (node1, node2) res;

  analog begin
    I(res) <+ V(res)/R;
  end

endmodule
```

We can model other components in a similar way. For example, a capacitor can be modeled as follows:

```
`include "disciplines.vams"
module capacitor (node1, node2);
  inout node1, node2;
  electrical node1, node2;
  parameter real C = 1;
  branch (node1, node2) cap;

  analog begin
    I(cap) <+ C*ddt(V(cap));
  end

endmodule
```

The **ddt** function evaluates the time derivative. Similarly, there is an **idt** function for calculating the time integral. Because the contribution statement is an algebraic expression and not an assignment, it is also possible to write the capacitor equation as:

```
V(cap) <+ idt(I(cap))/C;
```

Before leaving these basic models, let us consider a pure voltage source that generates a sine wave. We will need a version of this element to describe a DAC.

```
`include "constants.vams"
`include "disciplines.vams"

module vsin(a,b);
inout  a,b;
electrical a,b;
branch(a,b)  vs;
parameter  real  vo = 1;
parameter real va = 1;
parameter real freq = 1;

analog  begin
  V(vs) <+ vo + va * sin(`M_TWO_PI*freq*$abstime);
end
endmodule
```

The file constants.vams defines a number of useful constants, including M_TWO_PI – 2π. **$abstime** is the current simulation time. This is analogous to the simulation time in "standard" Verilog—**$time**—but is a real number.

14.3.3 Mixed-Signal Modeling

Verilog-AMS is a mixed-signal modeling language. Therefore, we can mix analog and digital constructs in the same models. Let us consider a simple comparator. We want to convert two analog voltages into a 1-bit digital signal, such that the output is a logic 1 when the the first input is greater than the second and 0 otherwise. The model is written as:

```
`include "disciplines.vams"
module comp(Aplus, Aminus, Dout);
  inout Aplus, Aminus;
  electrical Aplus, Aminus;
  output Dout;
  reg Dout;
```

```
  initial
    begin
    Dout = 1'b1;
    forever
      begin
      @(cross(V(Aplus, Aminus), -1)) Dout = 1'b0;
      @(cross(V(Aplus, Aminus), +1)) Dout = 1'b1;
      end
    end

endmodule
```

This module has three ports—two are electrical nodes and one is a digital net. In the module body, we need to detect when one voltage becomes greater or less than the other and switch the output accordingly. This could be done with a simple comparison operator, but it is better to use the **cross** function. When the expression crosses zero, an event is created for the digital signal. A second parameter is used to specify the direction if only one crossing direction should trigger the event— +1 for positive, –1 for negative, and 0 or unspecified for either direction. The **cross** function, however, does not trigger an event for the initial conditions. Thus, an initial block is created to give Dout an initial value and the rising and falling transitions are tested to generate events and to change Dout.

This example simply converts a signal to 1 bit. We can use the comparator as part of a flash ADC (see Section 14.2 and Exercise 14.5). Later, we will use the comparator again as part of a tracking ADC. We can, however, also model a flash ADC behaviorally. We simply need to convert a varying (real) quantity into a bit vector. In the example that follows, the model is parameterized in terms of the analog voltage range and the number of bits. We also include a clock to sample the waveform—otherwise, the model will be evaluated at every analog time step. `Dmax` is set to all 1s and then converted to a real number to scale the result.

```
`include "disciplines.vams"
module adc(Dout, Ain, clock);
  parameter N = 8;
  parameter real Vrange;
  output [N-1:0] Dout;
  reg [N-1:0] Dout;
  inout Ain;
  electrical Ain;
  input clock;

  parameter [N-1:0] Dmax = {N{1'b1}};
```

```
  always @(posedge clock)
    Dout <= V(Ain)/Vrange*real(Dmax);

endmodule
```

In the following example, a DAC is modeled as a voltage source and resistance in the analog world. The voltage source can take one of three values—V1, V0, or Vx for logic 1, logic 0, or unknown, respectively. Similarly, the output resistance can take a low impedance value or a high impedance value.

To convert from analog quantities to digital signals, we write procedural statements. To convert the other way, we need to write contribution statements. There is a catch, however. In discrete simulation ("standard" Verilog), signals change instantaneously. In a continuous simulation, instantaneous step changes cause problems.

Without going into great detail, an analog or continuous solver approximates a changing quantity by taking discrete time steps. The waveform is therefore approximated by a polynomial expression. The size of these time steps is varied to minimize the error in the polynomial. A large step change makes it impossible to construct a polynomial expression across that change, so the error is considered large and the time step is reduced in an attempt to minimize the error. No matter how small the time step is made, the error will remain large and the simulation fails.

One way to avoid instantaneous changes is to force a transition to occur in a finite time. This can be done with the **transition** function. The values of the voltage and resistance are held as variables within the analog block of the DAC model. When the input signal changes, these variables are updated. An expression for the output voltage in terms of these signals can then be written as a contribution statement. Note that changes in the signals are slowed by 1 ns (expressed as a real number) using the **transition** function.

```
`include "disciplines.vams"

module dac(Din, Aout);

input Din;
inout Aout;
electrical Aout;

parameter real V1 = 5.0;
parameter real V0 = 0.0;
parameter real Vx = 2.5;
parameter real Zhi = 1e9;
parameter real Zlo = 1;

real Zth, Vth;
```

```
analog begin
  Zth = (Din === 1'bz) ? Zhi: Zlo;
  Vth = (Din === 1'b0) ? V0 :
        (Din === 1'b1) ? V1 : Vx;
  V(Aout) <+ transition(Vth, 1e-9)
              - I(Aout)*transition(Zth, 1e-9);
  end
endmodule
```

The obvious disadvantage of this approach is that the time to change between values has to be specified. 1 ns might easily be far too large or far too small compared with other changes in the system. It would be better to let the solver decide for itself what would constitute a suitable change. For this to happen, the solver needs to be told that there could be a problem, and this is the responsibility of the model writer. Verilog-AMS includes a mechanism for indicating a discontinuity—the **$discontinuity** function.

```
`include "disciplines.vams"

module dac(Din, Aout);

input Din;
inout Aout;
electrical Aout;

parameter real V1 = 5.0;
parameter real V0 = 0.0;
parameter real Vx = 2.5;
parameter real Zhi = 1e9;
parameter real Zlo = 1;

real Zth, Vth;

analog begin
  @(Din) $discontinuity;
  case (Din)
    1'b0 : V(Aout) <+ V0 - I(Aout) * Zlo;
    1'b1 : V(Aout) <+ V1 - I(Aout) * Zlo;
    1'bz : V(Aout) <+ Vx - I(Aout) * Zhi;
    1'bx : V(Aout) <+ Vx - I(Aout) * Zlo;
    endcase
  end

endmodule
```

When Din changes, the analog solver stops and restarts, therefore avoiding the error detection mechanism. Each branch of the case statement consists of one contribution statement, modeling a Thévenin equivalent circuit.

If we wish to convert several bits to an analog equivalent, we could use a 1-bit DAC for each input bit and add the outputs together, with appropriate weighting. If we are not concerned with converting X and Z bits, it is easier to simply convert the bits to a real number as follows. Notice that the output is scaled to a parameter, Vref.

```
'include "disciplines.vams"
module NbitDac(Din, Aout);
  parameter N = 8;
  parameter real Vref = 1.0;
  input [N-1:0] Din;
  inout Aout;
  electrical Aout;

  analog
    V(Aout) <+ Din*Vref/((1<<N)-1);

endmodule
```

We now have the necessary parts to build the tracking ADC from Section 14.2. We also need the counter from Exercise 5.5. This has been written as a self-contained testbench. Notice that we have created a netlist in exactly the same way as a digital netlist, the only difference being that the analog nodes needed for connecting components are declared as electrical nodes.

```
'include "disciplines.vams"
'timescale 1 ns / 100 ps

module tracking;

electrical Ain, Aout, gnd;
ground gnd;
wire Up;
wire [3:0] Dout;

reg Clock, Reset;

initial
  begin
  Clock = 1'b0;
  forever
    #10 Clock = ~Clock;
  end
```

```
initial
  begin
     Reset = 1'b1;
  #2 Reset = 1'b0;
  end

UpDown #(.N(4)) C1 (.clk(Clock), .reset(Reset),
                     .up(Up), .Count(Dout));
NBitDac #(.N(4), .Vref(5.0)) D1
         (.Din(Dout), .Aout(Aout));
comp O1 (.Aplus(Ain), .Aminus(Aout), .Dout(Up));
vsin #(.vo(2.5), .va(2.5), .freq(1e4)) V1
      (.a(Ain), .b(gnd));

endmodule
```

14.4 Phased-Locked Loops

Although ADCs and DACs are the main interfaces between the analog and digital worlds, another class of circuits also sits at this boundary. One of the major uses for phase-locked loops (PLLs) is for generating clocks. PLLs can be used to recover the clock from a stream of data. A PLL can also be used to "clean up" a clock that has an irregular period and to multiply a clock signal to create a higher frequency signal. All of these tasks are difficult to achieve with conventional digital circuit techniques. PLLs can be built as purely analog circuits, purely digital circuits, or using a mixture of methods. As with ADCs and DACs, there is not enough space in a book like this to give any more than a brief introduction to PLLs. The purpose here is to show a simple example and to illustrate one way of modeling that example in Verilog-AMS. As with ADCs and DACs, the real art of modeling PLLs is to capture non-linearities and other imperfections to determine whether a particular design will work in a particular context.

Figure 14.6 shows the basic structure of a PLL. The phase detector determines the difference between the input (ref_clk) and the stabilized output (vco_out). The phase detector could be an analog four-quadrant multiplier, a digital XOR gate, or a sequential digital circuit. The output from the phase detector is a sequence of pulses. The low pass filter averages these pulses in time. This filter is crucially important to the working of the PLL. If the time constant is too small, the PLL will not settle into a regular "locked" pattern. If the time constant is too great, the PLL may not lock at all. The voltage controlled oscillator (VCO) converts the output of the filter into a oscillation whose frequency is determined by the filter output voltage. The VCO is likely to be the hardest part of the design. It can only oscillate within a relatively

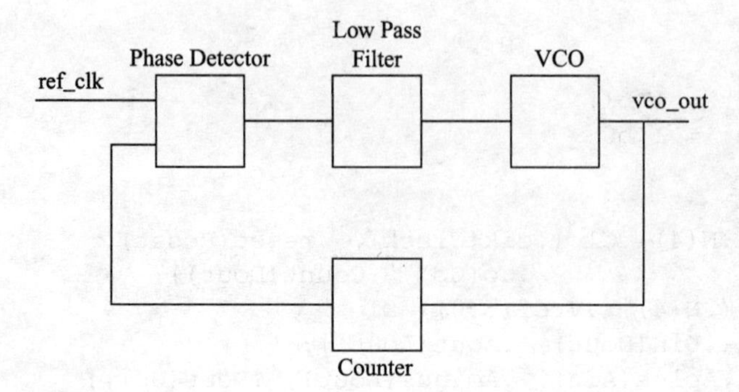

Figure 14.6 PLL structure.

narrow band of frequencies. Finally, the counter is optional. By dividing the VCO output, the phase detector compares with this reduced frequency output. In other words, the VCO output must be a multiple of the input frequency.

There are many books about PLL design, but perhaps the best way to understand their operation is by playing with the circuit parameters in a simulation. Therefore, we will simply present one, ideal, model of a PLL.

We start with the largest model—the phase detector. We will use the example from Chapter 13. The two outputs, qa and qb, correspond to two control signals, up and down, respectively. These need to be converted to analog voltages and filtered. We will use two instances of the 1-bit DAC from the previous section. The low pass filter can be modeled using the Laplace transform attribute in Verilog-AMS. This attribute takes two parameters, each of which is a vector of real numbers. The first vector contains the coefficients of the numerator and the second contains the coefficients of the denominator. Here we want to create a parameterized low pass filter, which in the s-domain has the transfer function:

$$\frac{1}{1+sT}.$$

Therefore, the numerator has the value 1.0, and the denominator has the values 1.0 and T. Hence, this is the Verilog-AMS model. Although this is a frequency domain model, it can be interpreted in the time domain. Similarly, time domain models (such as **ddt**(v)) can be interpreted in the frequency domain.

```
`include "disciplines.vams"

module lpf (Ao, Ai);
```

```
inout Ao, Ai;
electrical Ao, Ai;

parameter real T = 1e-6 from [0:inf);

analog
  V(Ao) <+ laplace_nd(V(Ai), {1}, {1, T});

endmodule
```

Notice that the range of the parameter, T, is limited to positive real numbers by using the **from** modifier. The mixture of square and round brackets means that 0 can be included in the range of possible numbers, but that infinity is excluded.

The voltage controlled oscillator is mixed-signal, but can be written using a Verilog process.

```
`include "disciplines.vams"
`timescale 1 s / 100 ps

module vco(Ina, Inb, vout);

inout Ina, Inb;
output vout;
electrical Ina, Inb;
reg vout;

parameter real gain = 5e5;
parameter real fnom = 2.5e5;
parameter real vc = 2.5;

real frequency;
real period;

always
  begin
  frequency = fnom + (V(Ina) - V(Inb) - vc) * gain;
  if (frequency > 0.0)
          period = 1/frequency;
  else
    period = 1/fnom;
  #(period/2) vout = 1'b1;
  #(period/2) vout = 1'b0;
  end

endmodule
```

Note that there are Verilog-AMS extensions, even within the always block. The values of the input quantities are found using the V() function call and the wait statements take real numbers, not time units.

The counter is purely digital, although we will use it in an asynchronous way. This is a SystemVerilog model—we are assuming the simulator can accept a mixture of languages and versions.

```
module counter #(parameter N = 4)
                (output reg count,
                 input clk);

integer cnt = 0;

always_ff @(posedge clk)
    begin
    cnt++;
    if (cnt == N)
      begin
      cnt = 0;
      count <= 1'b1;
      end
    else
      count <= 1'b0;
    end

endmodule
```

Finally, we can put all the parts together and include a suitable stimulus.

```
`include "disciplines.vams"
`timescale 1 ns / 100 ps

module pll;

electrical up, down, up_a, down_a;
wire up_d, down_d, VCO_out, VCO_div;

reg ref_clk;

initial
  begin
  ref_clk = 1'b0;
  forever
    #5000 ref_clk = ~ref_clk;
  end
```

```
phase_detector P0 (.inA(ref_clk), .inB(VCO_div),
                   .qA(up_d), .qB(down_d));
dac D0 (.Din(up_d), .Aout(up_a));
dac D1 (.Din(down_d), .Aout(down_A));
lpf #(50e-6) L0 (.Ai(up_a), .Ao(up));
lpf #(50e-6) L1 (.Ai(down_a), .Ao(down));
vco #(.gain(1e5), .fnom(8e5), .vc(2.5)) V0
    (.Ina(down), .Inb(up), .vout(VCO_out));
counter #(5) C0 (.clk(VCO_out), .count(VCO_div));

endmodule
```

Simulation of this PLL model shows that the output frequency varies between about 450kHz and 600kHz, before settling at 500kHz after about 260 μs. The clock has a frequency of 100kHz and the counter counts to 5, so the PLL behaves exactly as we would expect.

14.5 Verilog-AMS Simulators

It could be argued that the mixed-signal models of ADCs, DACs, and PLLs could be modeled entirely in standard Verilog. Indeed, there is a very limited amount of behavior that requires an analog solver in these models. The real power of Verilog-AMS is that it allows digital Verilog models to be simulated at the same time as analog circuits that would traditionally have been simulated with SPICE. It is possible to include SPICE netlists within Verilog-AMS. Many simulators are now multi-language. It is therefore possible to have models written in combinations of the various dialects of Verilog and SystemVerilog, Verilog-AMS, SPICE, VHDL, VHDL-AMS, and SystemC.

Summary

At some point, digital circuits have to interface with the real, analog world. Modeling this interface and the interaction with analog components has always been difficult. Verilog-AMS extends Verilog to allow analog and mixed-signal modeling. Typical converters include ladder DACs, flash ADCs, delta-sigma ADCs, and PLLs. All of these components can be modeled and simulated in Verilog-AMS. There is, as yet, no way to automatically synthesize such elements from a behavioral description. Verilog-AMS simulators are still relatively new and may not support the entire language. They do, however, provide means for interfacing between SPICE models and Verilog-AMS, allowing modeling of complete systems.

Further Reading

For an explanation of analog simulation algorithms, see Litovski and Zwolinski [12]. Horowitz and Hill [31] is an excellent guide to practical circuit design and includes descriptions of ADCs, DACs, and PLLs. For a full description of Verilog-AMS, the LRM is, of course, invaluable. Manufacturers' manuals need to be read with the LRM to understand any limitations.

Exercises

14.1 An inductor is described by the equation $v_L = L \cdot \frac{di_L}{dt}$. Write a Verilog-AMS model of an inductor, using the **`ddt`** function.

14.2 Write an inductor model that uses the **`idt`** function.

14.3 Write a parameterizable model of a voltage source that generates a ramp. The parameters should be initial voltage, final voltage, delay before the ramp, and rise (or fall) time.

14.4 Write a model of voltage source that generates a pulse. What parameters need to be specified? How is it made to repeat?

14.5 Write a Verilog-AMS model of the flash ADC shown in Figure 14.1.

A SystemVerilog and Verilog

A.1 Standards

The current standard of SystemVerilog is published by the IEEE and Accellera, an industry-based standards association. It is given the number 3.1a, which indicates that it is the third major revision of Verilog. These are the SystemVerilog and Verilog standards.

IEEE Standard 1800-2005/IEC 62530:2007

This is the LRM. This, however, only defines the *extensions* to the Verilog standard—1364-2005. To get a full definition of SystemVerilog, both LRMs are needed.

IEEE Standard 1364-2005

This is the latest standard for Verilog. There was also a 2001 version, but the changes are relatively minor.

IEEE Standard 1364-1995

The original Verilog standard. Most of the open source tools conform only to this standard.

IEEE Standard 1364.1-2002/IEC 62142-2005

Verilog RTL synthesis standard. This defines what is and what is not synthesizable. It is based on the 2001 Verilog standard.

Accellera Verilog-AMS 2.3-2008

Verilog-AMS. This is based on the 2005 Verilog standard. There is also a Verilog-A subset that includes only the continuous time elements. It is an ambition of Accellera to create SystemVerilog-AMS.

A.2 SystemVerilog and Verilog Differences

There are, of course, a very large number of additions to the basic syntax of Verilog in SystemVerilog. Features such as assertions, classes, and programs have been added. The purpose of this section is to show changes to the two older versions of Verilog that affect RTL hardware modeling and hence to show how the examples in this book might be modified to work with tools that support only these earlier standards. As an illustration, we will use the state machine modeling the traffic light controller of Section 6.5.1.

Verilog 2005

The Verilog 2005 version of the state machine is shown in the following paragraphs. SystemVerilog is a superset of Verilog 2005, so this is valid SystemVerilog code. The important changes from the SystemVerilog version are:

- **always_ff** is replaced by **always**. Arguably, **always_ff** is redundant because the meaning and syntax of the two forms are identical.
- **always_comb** is replaced by **always** `@(*)`. The sensitivity list is a wild card, which means that any variables or nets used as inputs to the block are automatically included.
- **logic** is replaced by **reg**. The two forms are interchangeable in SystemVerilog, but **logic** clearly means a logic type, whereas **reg** might be mistakenly assumed to refer to a register.
- There are no enumerated types in Verilog. The construct is faked in Verilog by declaring a variable with sufficient bits to hold all the states and then declaring **parameter** values for each state. Here, two states can be represented by a single bit. Four states, for example, would be declared using

reg [1:0]. The clear disadvantage of this approach is that the state assignment is done at the same time that the state machine is designed.

- Bit strings of undefined length, such as '0 are not permitted. The string length must be specified, for example, 1'b0.
- SystemVerilog makes no distinction between variables and nets in assignments. In Verilog, variables must be assigned values in procedures; nets (wires) must only be assigned values in continuous assignments.

```
module traffic_1_05 (output reg start_timer,
                     major_green, minor_green,
                     input reg clock, n_reset, timed,
                     car);
  reg state;
  parameter G=0, R=1;

always @(posedge clock, negedge n_reset)
  begin: SEQ
  if (~n_reset)
    state <= G;
  else
    case (state)
      G: if (car)
           state <= R;
      R: if (timed)
           state <= G;
    endcase
  end


always @(*)
  begin: OP
  start_timer = 1'b0;
  minor_green = 1'b0;
  major_green = 1'b0;
  case (state)
    G: begin
       major_green = 1'b1;
       if (car)
         start_timer = 1'b1;
       end
    R: minor_green = 1'b1;
  endcase
  end

endmodule
```

Verilog 1995

The Verilog 1995 version of the state machine follows. SystemVerilog and Verilog 2005 are both supersets of Verilog 1995, so this is valid code for both later versions. The important changes from the Verilog 2005 version are:

- So-called ANSI C-style module headers are not permitted. Instead, the input and output signal names only are written in the header. Then, the signal modes must be declared: **input**, **output**, or **inout**. Finally, any outputs of type **reg** must be declared—the default is **wire**. Thus, outputs may be listed three times.
- The separator between items in sensitivity or event lists must be **or**, not a comma. This is not a logical OR of Boolean values, but an OR of events.
- Wild cards are not permitted in sensitivity lists. Every input to the combinatorial process must be listed. "Inputs" include anything on the right-hand side of an assignment or anything in the decision part of an **if** or a **case** statement. This is a source of potential significant error. If an input is missing, the process will not be simulated correctly, but a synthesis tool is likely to "correct" the mistake. Thus, there would be a mismatch between synthesis and simulation.

```
module traffic_1_95 (start_timer, major_green,
                     minor_green, clock, n_reset,
                     timed, car);

  output start_timer, major_green, minor_green;
  input clock, n_reset, timed, car;

  reg start_timer, major_green, minor_green;
  reg state;
  parameter G=0, R=1;

always @(posedge clock or negedge n_reset)
  begin: SEQ
  if (~n_reset)
    state <= G;
  else
    case (state)
      G: if (car)
           state <= R;
```

```
      R: if (timed)
           state <= G;
    endcase
  end


always @(timed or car or present_state)
  begin: OP
  start_timer = 1'b0;
  minor_green = 1'b0;
  major_green = 1'b0;
  case (state)
    G: begin
       major_green = 1'b1;
       if (car)
         start_timer = 1'b1;
       end
    R: minor_green = 1'b1;
  endcase
  end

endmodule
```

Answers to Selected Exercises

Exercise 3.1

```
module Nand3 (output wire z, input wire w, x, y);

assign #5ps z = ~(w & x & y);

endmodule
```

Exercise 3.3

```
module Full_Adder(output S, Co, input a, b, Ci);

  wire na, nb, nc, d, e, f, g, h, i, j;

  not n0 (na, a);
  not n1 (nb, b);
  not n2 (nc, Ci);
  and a0 (d, na, nb, Ci);
  and a1 (e, na, b, nc);
  and a2 (f, a, b, Ci);
  and a3 (g, a, nb, nc);
  or o0 (S, d, e, f, g);
  and a4 (h, b, Ci);
  and a5 (i, a, b);
  and a6 (j, a, Ci);
  or o1 (Co, h, i, j);

endmodule
```

Exercise 3.4

```
module TestAdder;

  wire a, b, Ci, S, Co;

  FullAdder f0 (.*);
```

```
  initial
    begin
    a = '0; b = '0; Ci = '0;
    #10ns a = '1;
    #10ns a = '0; b = '1;
    #10ns a = '1;
    #10ns a = '0; b = '0; Ci= '1;
    #10ns a = '1;
    #10ns a = '0; b = '1;
    #10ns a = '1;
    end

endmodule
```

Exercise 4.3

```
module Bool3to8(output logic [7:0] z,
                input logic [2:0] a);

  always_comb
    begin
    z[0] = ~a[0] & ~a[1] & ~a[2];
    z[1] = a[0] & ~a[1] & ~a[2];
    z[2] = ~a[0] & a[1] & ~a[2];
    z[3] = a[0] & a[1] & ~a[2];
    z[4] = ~a[0] & ~a[1] & a[2];
    z[5] = a[0] & ~a[1] & a[2];
    z[6] = ~a[0] & a[1] & a[2];
    z[7] = a[0] & a[1] & a[2];
    end

endmodule

module Cond3to8(output logic [7:0] z,
                input logic [2:0] a);

  always_comb
    z = (a == 3'b000) ? 8'b00000001 :
        (a == 3'b001) ? 8'b00000010 :
        (a == 3'b010) ? 8'b00000100 :
        (a == 3'b011) ? 8'b00001000 :
        (a == 3'b100) ? 8'b00010000 :
        (a == 3'b101) ? 8'b00100000 :
        (a == 3'b110) ? 8'b01000000 :
        (a == 3'b111) ? 8'b10000000 :
        'x;

endmodule
```

```
module Shif3to8(output logic [7:0] z,
                input logic [2:0] a);

  always_comb
    z = 1'b1 << a;

endmodule


module Test3to8;

logic [2:0] a;
logic [7:0] z0, z1, z2;

Bool3to8 d0 (z0, a);
Cond3to8 d1 (z1, a);
Shif3to8 d2 (z2, a);

initial
  begin
        a = 3'b000;
  #10ns a = 3'b001;
  #10ns a = 3'b010;
  end

endmodule
```

Exercise 4.4

```
module Priority #(parameter N = 3)
         (output logic [N-1:0] y,
          output logic valid,
          input logic [(1<<N)-1:0] a);

  always_comb
    begin
    valid = '0;
    y = '0;
    for (int i = N-1; i >= 0; i--)
      if (a[i])
        begin
        y = i;
        valid = '1;
        end
    end

endmodule
```

Exercise 4.6

```
module Comparator #(parameter N = 3)
         (output logic [N-1:0] eq,
          input logic [N-1:0] x, y);

logic eqi;

  always_comb
    begin
    eqi = '1;
    for (int i = 0; i < N; i++)
      eqi = ~(x[i] ^ y[i]) & eqi;
    eq = eqi;
    end

endmodule
```

Exercise 5.2

```
module dffrs (output logic q,
              input logic d, clk, reset, n_set);

always_ff @(negedge clk, negedge n_set)
  if (~n_set)
    q <= '1;
  else if (reset)
    q <= '0;
  else
    q <= d;

endmodule
```

Exercise 5.5

```
module counterud #(parameter N = 8)
                 (output logic [N-1:0] count,
                  input logic n_reset, clk, up);

always_ff @(posedge clk, negedge n_reset)
  if (~n_reset)
    count <= 0;
  else if (up && (count < ((1'b1<<N) - 1)))
    count <= count + 1;
  else if (!up && (count > 0))
    count <= count - 1;

endmodule
```

Exercise 5.6

```
module piso #(parameter N = 8) (output logic q,
              input logic [N-1:0] a, input logic clk, load);

logic [N-1:0] qr;

always_ff @(posedge clk)
  if (load)
    qr <= a;
  else
    qr <= {1'b0, qr[N-1:1]};

always_comb
  q = qr[0];

endmodule
```

Exercise 5.11

```
module countlfsr (output logic [2:0] count,
                  input logic n_reset, clk);

always_ff @(posedge clk, negedge n_reset)
  if (~n_reset)
    count <= '1;
  else
    count <= {count[1:0], count[2] ^ count[1]};

endmodule
```

Exercise 6.3

```
always_ff @(posedge clock, negedge n_reset)
  begin: SEQ
  if (~n_reset)
    present_state <= S0;
  else
    present_state <= next_state;
  end
```

Exercise 6.5

```
module seqdet (output logic z,
               input logic clock, n_reset, x);

  enum {s0, s1, s2} state;
```

```
always_ff @(posedge clock, negedge n_reset)
  begin: SEQ
  if (~n_reset)
    state <= s0;
  else
    case (state)
      s0: if (~x)
             state <= s0;
           else
             state <= s1;
      s1: if (~x)
             state <= s0;
           else
             state <= s2;
      s2: if (~x)
             state <= s0;
           else
             state <= s2;
    endcase
  end

always_comb
  begin: COM
  if (state == s2 && x)
    z = '1;
  else
    z = '0;
  end

endmodule
```

Exercise 6.8

```
module twoseq_1 (output logic z,
                 input logic clock, n_reset, a, b);

  enum {s0, s1, s2} state;

always_ff @(posedge clock, negedge n_reset)
  begin: SEQ
  if (~n_reset)
    state <= s0;
  else
    case (state)
      s0: if (a && b)
             state <= s1;
           else
```

```
            state <= s0;
      s1: if (a && !b)
            state <= s2;
          else
            state <= s0;
      s2: state <= s0;
    endcase
  end

always_comb
  begin: COM
  if (state == s2 && !a && !b)
    z = '1;
  else
    z = '0;
  end

endmodule
```

Exercise 6.9

```
module twoseq_2 (output logic z,
                 input logic clock, n_reset, a, b);

enum {s0, s1, s2} present_state, next_state;

always_ff @(posedge clock, negedge n_reset)
  begin: SEQ
  if (~n_reset)
    present_state <= S0;
  else
    present_state <= next_state;
  end

always_comb
  begin: COM
    z = '0;
    case (state)
      s0: if (a && b)
            next_state <= s1;
          else
            next_state <= s0;
      s1: if (a && !b)
            next_state <= s2;
          else
            next_state <= s0;
```

```
        s2: begin
            if (!a && !b)
              z = '1;
            next_state <= s0;
            end
      endcase
    end

endmodule
```

Exercise 7.6

State s9 is modified to load the PC from the Addr part of the IR.

```
    s9: begin
          bus.Addr_bus = '1;
          bus.load_PC = '1;
          end
```

In fact, this could be done in the same clock cycle as s6; thus, the sequential part is changed to:

```
s6: if (bus.op == LOAD)
      state <= s7;
    else if (bus.op == BNE)
      state <= s0;
    else
      state <= s8;
```

and the combinational part becomes:

```
      s6: begin
            bus.CS = '1;
            bus.R_NW = '1;
            if ((bus.op == BNE) && (!bus.z_flag))
              begin
              bus.Addr_bus = '1;
              bus.load_PC = '1;
              end
            end
```

Exercise 8.4

This is a liveness property.

The following is better because it tests the function and because it can be falsified after a finite period of time.

```
assert property (@(posedge clk) load |=> ##N ready);

assert property (@(posedge clk) load |-> ready);
```

Exercise 9.4

```
always_ff @(posedge clock, negedge n_reset)
  if (!n_reset)
    state <= s0;
  else
    case (state)
      s0: if (start)
              state <= s1;
      s1: if (timed)
              state <= s0;
    endcase

always_comb
  if (state == s1)
   enable = '1;
  else
    enable = '0;

always_ff @(posedge clock, negedge n_reset)
  if (!n_reset)
    count <= 0;
  else
    begin
    if (enable)
      count <= count + 1;
    if (count == 255)
      count <= 0;
    end

always_comb
  if (count == 255)
    timed = '1;
  else
    timed = '0;
```

Exercise 10.4

```
module fsm (input logic clk, a, reset, output logic y);

  typedef enum {s0, s1, s2} statetype;
  statetype currentstate, nextstate;

always_ff @(posedge clock, posedge reset)
  if (reset)
    currentstate <= s0;
```

```
  else
    currentstate <= nextstate;

always_comb // or always @(*) or always @(currentstate or a)
  begin
  y = '0;
  case (currentstate)
    s0: if (a)
          nextstate = s1;
        else
          nextstate = s2;
    s1: begin
        y = '1;
        nextstate = s0;
        end
    s2: if (a)
          nextstate = s2;
        else
          nextstate = s0;
    endcase
  end

endmodule
```

Exercise 11.3

Test for A/0: 0100/0, also covers E/1, G/1, H/0, I/0, J/1. Test for A/1: 1100/1, also covers B/0, C/1, D/1, E/0, F/0, H/1, J/0. Test for G/0 implies G = 1, hence B = C = 1. To propagate G to I implies F = 1, which implies C = D = 0. Hence, there is a contradiction.

Exercise 11.7

11..11/0, 11..10/1, 11..01/1, .., 10..11/1, 01..11/1

Exercise 12.3

50 flip-flops implies $2^{50} \approx 10^{15}$ states. At 1MHz it would take 10^9 sec ≈ 36 years to reach all states.

It takes 50 clock cycles to load the scan path (unloading can be done at the same time as loading the next pattern). 200 patterns take 10,000 cycles = 10 ms at 1Mhz.

Exercise 12.7

State	TMS	TDI
Test-Logic-Reset	0	—
Run-Test/Idle	1	—
Select-DR-Scan	1	—
Select-IR-Scan	0	—
Capture-IR	0	—
Shift-IR	0	0
Shift-IR	1	1
Exit1-IR	1	—
Update-IR	1	—
Select-DR-Scan	0	—
Capture-DR	0	—
Shift-DR	0	0
Shift-DR	0	1
Shift-DR	0	0
Shift-DR	1	1
Exit1-DR	1	—
Update-DR	0	—
Run-Test/Idle		

— means don't care. Change of state occurs on rising edge of TCK.

Exercise 12.11

```
module tap_controller (input logic tms, tck,
                       output logic ShiftIR, ClockIR, UpdateIR,
                       ShiftDR, ClockDR, UpdateDR);

enum {test_logic_reset, run_test_idle,
      select_DR_scan, capture_DR, shift_DR, exit1_DR,
      pause_DR, exit2_DR, update_DR,
      select_IR_scan, capture_IR, shift_IR, exit1_IR,
      pause_IR, exit2_IR, update_IR}
      present_state, next_state;

always_ff @(posedge tck)
  present_state <= next_state;
```

```
always_comb
  begin: COM
  ShiftIR = '0;
  ClockIR = '0;
  UpdateIR = '0;
  ShiftDR = '0;
  ClockDR = '0;
  UpdateDR = '0;
  case (present_state)
    test_logic_reset:
      begin
      if (!tms)
        next_state = run_test_idle;
      else
        next_state = test_logic_reset;
      end
    run_test_idle:
      begin
      if (tms)
        next_state = select_DR_scan;
      else
        next_state = run_test_idle;
        end
    select_DR_scan:
        begin
        if (tms)
          next_state = select_IR_scan;
        else
          next_state = capture_DR;
        end
    capture_DR:
        begin
        ClockDR = '1;
        if (tms)
          next_state = exit1_DR;
        else
          next_state = shift_DR;
        end
    shift_DR:
        begin
        ClockDR = '1;
        ShiftDR = '1;
        if (tms)
          next_state = exit1_DR;
        else
          next_state = shift_DR;
```

```
    end
  exit1_DR:
    begin
    if (tms)
      next_state = update_DR;
    else
      next_state = pause_DR;
    end
  pause_DR:
    begin
    if (tms)
      next_state = exit2_DR;
    else
      next_state = pause_DR;
    end
  exit2_DR:
    begin
    if (tms)
      next_state = update_DR;
    else
      next_state = shift_DR;
    end
  update_DR:
    begin
    UpdateDR = '1;
    if (tms)
      next_state = select_DR_scan;
    else
      next_state = run_test_idle;
    end
  select_IR_scan:
    begin
    if (tms)
      next_state = test_logic_reset;
    else
      next_state = capture_IR;
    end
  capture_IR:
    begin
    ClockIR = '1;
    if (tms)
      next_state = exit1_IR;
    else
      next_state = shift_IR;
    end
  shift_IR:
```

```
        begin
        ClockIR = '1;
        ShiftIR = '1';
        if (tms)
          next_state = exit1_IR;
        else
          next_state = shift_IR;
        end
      exit1_IR:
        begin
        if (tms)
          next_state = update_IR;
        else
          next_state = pause_IR;
        end
      pause_IR:
        begin
        if (tms)
          next_state = exit2_IR;
        else
          next_state = pause_IR;
        end
      exit2_IR:
        begin
        if (tms)
          next_state = update_IR;
        else
          next_state = shift_IR;
        end
      update_IR:
        begin
        UpdateIR = '1;
        if (tms)
          next_state = select_IR_scan;
        else
          next_state = run_test_idle;
        end
    endcase
  end

endmodule
```

Exercise 13.6

States A, E, and F can be merged. States B and C can be merged. An extra state (T) needs to be introduced to avoid races—let this be between D and AEF. A possible state assignment is AEF (00), BC (01), D (11), T (10), giving next state and output equations:

$$Y_1^+ = Y_1 \cdot Y_0 + \bar{P} \cdot R \cdot Y_0$$
$$Y_0^+ = P \cdot \bar{R} + \bar{P} \cdot R \cdot Y_0 + \bar{P} \cdot \bar{Y}_1 \cdot Y0$$
$$Q = Y_1$$

Exercise 13.7

There are three feedback loops in Figure 13.4. Insert a virtual buffer at A, (Y_1), between F and the NAND gate with output B (Y_2), and at Q (Y_3).

$$Y_1^+ = D \cdot R \cdot Y_2 + \bar{S} + Y_1 \cdot R \cdot C$$
$$Y_2^+ = Y_1 \cdot R + \bar{C} + Y_2 \cdot D \cdot R$$
$$Y_3^+ = Y_3 \cdot Y_1 \cdot R + Y_3 \cdot R \cdot \bar{C} + Y_3 \cdot Y_2 \cdot D \cdot R + Y_1 \cdot R \cdot C + \bar{S}$$

Exercise 14.1

```
`include "disciplines.vams"
module inductor (node1, node2);
  inout node1, node2;
  electrical node1, node2;
  parameter real L = 1;
  branch (node1, node2) ind;

  analog begin
    V(ind) <+ L*ddt(I(ind));
  end

endmodule
```

Exercise 14.3

```
`include  "disciplines.vams"

module  vramp(a,b);
inout  a,b;
electrical a,b;
branch(a,b)  vr;
parameter  real  vl = 0;
parameter real vh = 1;
parameter real td = 1;
parameter real tr = 1;
```

```
analog begin
  if ($abstime < td)
    V(vr) <+ vl;
  else if ($abstime < td + tr)
    V(vr) <+ vl + ($abstime-td)*(vh - vl)/tr;
  else
    V(vr) <+ vh;
  end
endmodule
```

Bibliography

[1] *HP Boundary-Scan Tutorial and BSDL Reference Guide*. Hewlett-Packard Company, 1990.

[2] *Standard for SystemVerilog—Unified Hardware Design, Specification, and Verification Language*. IEEE 1800–2005/IEC 62530:2007 (E), 2007.

[3] M. Abramovici, M.A. Breuer, and A.D. Friedman. *Digital System Testing and Testable Design (Revised Printing)*. IEEE Press, 1990.

[4] J. Bergeron. *Writing Testbenches Using SystemVerilog*. Springer-Verlag, New York, NY, 2006.

[5] S. Brown and Z. Vranesic. *Fundamentals of Digital Logic with Verilog Design*, 2nd ed., McGraw-Hill, New York, NY, 2007.

[6] G. de Micheli. *Synthesis and Optimization of Digital Circuits*. McGraw-Hill, New York, NY, 1994.

[7] M.D. Edwards. *Automatic Logic Synthesis Techniques for Digital Systems*. MacMillan Press, New York, NY, 1992.

[8] R.W. Hamming. *Coding and Information Theory*. Prentice-Hall, Englewood Cliffs, NJ, 1980.

[9] J.L. Hennessy and D.A. Patterson. *Computer Architecture a Quantitative Approach*. Morgan Kaufman Publishers, San Mateo, CA, 1990.

[10] F.J. Hill and G.R. Peterson. *Computer Aided Logical Design with Emphasis on VLSI*, 4th ed., John Wiley & Sons, New York, NY, 1993.

[11] K. Kundert and O. Zinke. *The Designer's Guide to Verilog-AMS*. Kluwer Academic Publishers, Boston, MA, 2004.

[12] V. Litovski and M. Zwolinski. *VLSI Circuit Simulation and Optimization*. Chapman & Hall, London, 1997.

[13] A.B. Maccabe. *Computer Systems: Architecture, Organization and Programming*. Richard D. Irwin, Homewood, IL, 1993.

[14] C. Maunder. *The Board Designer, Guide to Testable Logic Circuits*. Addison-Wesley, Reading, MA, 1992.

[15] A. Miczo. *Digital Logic Testing and Simulation*. John Wiley & Sons, New York, NY, 1987.

[16] Z. Navabi. *VHDL Analysis and Modeling of Digital Systems*. McGraw-Hill, New York, NY, 1993.

[17] M.S. Nixon. *Introductory Digital Design: A_Programmable_Approach*. MacMillan Press, New York, NY, 1995.

[18] S. Palnitkar. *Verilog HDL: A Guide in Digital Design and Synthesis*, 2nd ed., Prentice Hall, Upper Saddle River, NJ, 2003.

[19] D.R. Smith and P.D. Franzon. *Verilog Styles for Synthesis of Digital Systems*. Prentice Hall, Upper Saddle River, NJ, 2000.

[20] D.J. Smith. *HDL Chip Design*. Doone Publishing, Madison, AL, 1996.

[21] C. Spear. *SystemVerilog for Verification: A Guide to Learning the Testbench Language Features*, 2nd ed., Springer-Verlag, New York, NY, 2008.

[22] S. Sutherland, S. Davidmann, and P. Flake. *SystemVerilog for Design: A Guide to Using SystemVerilog for Hardware Design and Modeling*, 2nd ed., Springer-Verlag, New York, NY, 2006.

[23] S.H. Unger. Hazards, critical races, and metastability. *IEEE Transactions on Computers*, 44(6):754–768, 1995.

[24] S. Vijayaraghavan and M. Ramanathan. *A Practical Guide for SystemVerilog Assertions*. Springer-Verlag, New York, 2005.

[25] J.F. Wakerley. *Digital Design Principles and Practices*, 2nd ed., Prentice Hall, Englewood Cliffs, NJ, 1994.

[26] N.H.E. Weste and K. Eshraghian. *Principles of CMOS VLSI Design: A Systems Perspective*, 2nd ed., Addison-Wesley, Reading, MA, 1992.

[27] M. Weyerer and G. Goldemund. *Testability of Electronic Circuits*. Carl Hanser Verlag, 1992.

[28] B.R. Wilkins. *Testing Digital Circuits*. Van Nostrand Reinhold, 1986.

[29] B. Wilkinson. *Digital System Design*, 2nd ed., Prentice Hall, Englewood Cliffs, NJ, 1992.

[30] W. Wolf. *Modern VLSI Design A Systems Approach*. Prentice Hall, Englewood Cliffs, NJ, 1994.

[31] P. Horowitz and W. Hill. *The Art of Electronics*. Cambridge University Press, New York, NY, 1989.

Index

Symbols and Numbers

A

Q

R

W

X